THE EARTH AS IT WAS

MICAH VAN HUSS

All Scripture quotations are from the King James Version of the Holy Bible.

Beacon Street Press

500 Beacon Drive
Oklahoma City, OK 73127

1-800-652-1144

www.swrc.com

Printed in the United States of America

ISBN: 978-1-933641-81-2

ACKNOWLEDGMENTS

For my wife Annie—who through hard work and love has built a home and family that enables my studies.

Thank you, dad, for discipline and for teaching me the way a man should go. You showed the love of Christ and provided for your family.

Thank you, Dr. Kent Hovind, for having the courage to teach the truth in the face of great adversity and persecution.

Thank you, Dr. Michael Heiser, for taking the Bible as the literal Word of God and teaching its truth without fear of man's scrutiny.

THE EARTH AS IT WAS

CONTENTS

PREFACE

But thou, O Daniel, shut up the words, and seal the book,
even to the time of the end: many shall run to and fro,
and knowledge shall be increased.– Daniel 12:4

I've been studying the mysteries of God's Word and the universe for decades, yet I'm amazed to learn about new things that I've never heard of before. This revelation of knowledge to our generation, I believe, is divine as Daniel 12 states. Are we approaching the time of the end?

Our understanding of the world in the year 2023 is trapped in the physical prison of our brains. If I ask you to think of a new color, can you do it? Is your brain capable of thinking of a new color?

We live in the bubble of today. So many things from scripture, history and mythology are too fantastical for most to give them a second thought. You picked up this book, so you're likely like me and are intrigued by the mysterious.

The world before the Flood of Noah was vastly different than it is now. In this book, I attempt to paint a picture of Elohim's antediluvian world. I will lay out as best I can why there is so much sin and suffering in our fallen world. I will attempt to prove that there is a whole lot more to mythologies than just fairy tales and fantasy.

More than any of that, I hope to point you to Jesus Christ. While it's awesome to study the mysteries of the universe, it's for naught if you miss the point of it all.

This book is full of my speculation about Elohim's universe. My speculations will be far stretched, fantastical, based somehow in scripture, but never against scripture. In other words, I keep an extremely open mind about the mysteries of our universe but never in contradiction to the Word of God.

I believe that angels rebelled against Elohim starting in the Garden of Eden. I believe that five generations after creation, those angels took on human form and took wives from among human women. I believe their offspring were giants and those giants' spirits are the demons that exist on the Earth today.

I believe that dinosaurs and man coexisted. I believe fire-breathing dragons existed and that heroes of old hunted them as trophies. I believe that some dinosaurs are still alive today.

I believe that on this planet existed, and may still exist, a tree that gives immortality to any person that eats its fruit. I believe that God allows us to see something that exists in the spiritual realm – something that you've seen.

I believe that ancient mankind possessed technologies that we have yet to discover. It's highly likely that nuclear weapons were used in warfare before the flood of Noah.

In this book, we cover all of these topics and more. I hope your encounter with *The Earth as it Was* is as fun for you to read as it was for me to write.

CHAPTER 1

ELOHIM'S CREATION

Before our six days of creation, Earth was an orb of water. Was ours the only one? How many planets with life has Elohim created across His universe? I hope that those who believe in God would not deny that He has the power to create life elsewhere. God has no beginning and no end. Are we His sole focus? We know that the spiritual and physical realms are separate. How many dimensions are there?

One can't deny that Mars, at one point, had water flowing on its surface. Was there life on Mars 7,000 years ago? It doesn't exclude any of my beliefs about God and His creation. It doesn't change what Jesus came to do for the Earth.

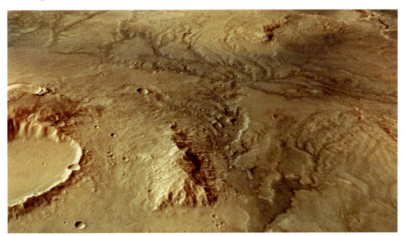

An image of ancient river valleys on Mars from ESA's Mars Express, Credit: European Space Agency, Wikimedia

Have other creations fallen? Do they have a similar salvation story? Because God preserved His Word for us here on Earth, these are questions about which we will likely not have an answer in this physical lifetime.

Notice in scripture that water was not established at creation. Water was already here before Elohim divides the light from the darkness, according to Genesis 1:1-3, *"In the beginning God created the heaven and the earth. And the earth was without form, and void; and darkness was upon the face of the deep. And the Spirit of God moved upon the face of the waters. And God said, Let there be light: and there was light."*

In our dimension (the physical realm), God had in place an orb of water. We know that God existed before this orb according to Proverbs 8:23-24, *"I was set up from everlasting, from the beginning, or ever the earth was. When there were no depths, I was brought forth; when there were no fountains abounding with water."*

God placed a layer of air in the orb of water to separate the water that is above from the water that is below. This idea is mentioned in 2 Peter 3:5, *"For this they willingly are ignorant of, that by the word of God the heavens were of old, and the earth standing out of the water and in the water:"*

Wouldn't it be fascinating if we ever discovered a planet that was only water? Well, we have. It's named GJ1214b and is 47 light years away from Earth. Is this one of Elohim's orbs waiting to be turned into a creation?

It appears that most, if not all of our angelic neighbors were created specifically for this Earth. After the six days of creation (yes, literal days), and before the seventh day, God says that the hosts were "finished." The Hebrew word for "host" is *ṣābā'* meaning army in Genesis 2:1, *"Thus the heavens and the earth were finished, and all the host of them."*

Were there angels created for places other than the Earth? Job 38 tells of angels that may have existed before our creation:

4Where wast thou when I laid the foundations of the earth? Declare, if thou hast understanding. 5Who hath laid the measures thereof, if thou knowest? Or who hath stretched the line upon it? 6Whereupon are the foundations thereof fastened? Or who laid the corner stone thereof; 7When the morning stars sang together, And all the sons of God shouted for joy?

Were these foundations that were being laid the orb of water that existed before the six days of creation? Angels were likely created before humans as scripture indicates they stand and shout at God's creation of the Earth.

Six Days of Creation

I believe God when He says in His word that He created the Earth in six days. Why? Because on the seventh day, Adam didn't rest for 1,000 years. Also, God says that He created the Earth in six days. Exodus 20:11 reads, *"for in six days the LORD made heaven and earth, the sea, and all that in them is, and rested the seventh day..."*

Because God gives us the ages of every man from Adam to Jesus, we know how old the Earth is. That is (from the current year, 2023) 6,197 years, according to the Masoretic Text and 7,577 years according to the Greek Septuagint.

There is a time gap in the Bible but it doesn't allow for billions of years of evolution. We see the gap between Genesis 1:1 and 2: *¹In the beginning God created the heaven and the earth. ²And the earth was without form, and void; and darkness was upon the face of the deep. – GAP – And the Spirit of God moved upon the face of the waters.*

This time gap is not compatible with evolution as there was only water. Verses 1 and 2a are not part of the first day of creation. In verse 2b God begins creation.

The only reason for a Christian to question the literal days of creation is in an attempt to believe the religion of evolution and its teachings of billions of years. It's a whole lot easier to believe that an intelligent powerful being created everything than to believe that nothing created everything.

No one likes authority. The first step in being free from authority is to deny God's existence. If God doesn't exist, then we are not accountable.

Secular scientists, geologists, professors, and the like are under a tremendous amount of pressure to deny God's existence. Most are threatened with excommunication from the religion of evolution thus ending their respective careers. Also, evolution comes from the same science that can't tell you what a woman is.

We see the same thing happening with today's culture. Bend the knee or cease to exist.

Evolution is a religious theory, not science. No one has ever observed one kind of animal change into another kind of animal. Animals and humans adapt to their environment. Adaptation is part of Elohim's creation. Darwin observed Elohim's adaptation of the Galapagos finches to formulate his theory.

Evolutionists' magic wand is time. Time is their explanation to try to get us to believe in the religion of Evolution. You've never observed evolution so you need faith to believe in it.

At the time of creation, there existed water and darkness. There are many types of creation theories. The Hebrew, along with Egyptian, Greek and Roman creation accounts are known as chaos creation theories. The watery abyss and the vast void are chaos.

On Day One, Elohim creates light. We read in Genesis 1:3-5, *"And God said, Let there be light: and there was light. And God saw the light, that it was good: and God divided the light from the darkness. And God called the light Day, and the darkness he called Night. And the evening and the morning were the first day."*

At this point, God has yet to create the sun. Was this a spiritual light – something in the spiritual realm? Can creation be considered an emanation from God without this light? Is this light a reflection of God's being, His divine nature or His personal will? 1 John 1:5 confirms this, *"This then is the message which we have heard of him, and declare unto you, that God is light, and in him is no darkness at all."*

On Day Two, Elohim creates the firmament in the midst of the waters. We read in Genesis 1:6-8, *"And God said, Let there be a firmament in the midst of the waters, and let it divide the waters from the waters. And God made the firmament, and divided the waters which were under the firmament from the waters which were above the firmament: and it was so. And God called the firmament Heaven. And the evening and the morning were the second day."*

You'll notice in the six days of creation, that the firmament created on Day Two is the only creation that God does not call good. Why was this?

The great expanse of water was one of the things that separated us from God. The firmament separated the Earth from the heavens. God of course passed through it and visited Earth many times in the antediluvian portion of the Old Testament. Maybe the separation was why God did not pronounce it good.

A more likely reason that Elohim did not pronounce the firmament good is the fact that it was not a permanent fixture of creation. The canopy above the firmament was created only to be broken 2,256 years later at the great flood of Noah in Genesis 7. We will further study the ice canopy in our chapter on the firmament.

On Day Three, Elohim creates the dry ground and plants. We read in Genesis 1:

> 9And God said, Let the waters under the heaven be gathered together unto one place, and let the dry land appear: and it was so. 10And God called the dry land

Earth; and the gathering together of the waters called he Seas: and God saw that it was good. ¹¹ And God said, Let the earth bring forth grass, the herb yielding seed, and the fruit tree yielding fruit after his kind, whose seed is in itself, upon the earth: and it was so. ¹² And the earth brought forth grass, and herb yielding seed after his kind, and the tree yielding fruit, whose seed was in itself, after his kind: and God saw that it was good. ¹³ And the evening and the morning were the third day.

This is the day that Pangea was formed. Today, we have seven continents (Eight if you believe that Zealandia is worthy of recognition). Pangea was the supercontinent that existed in the world before Noah's flood when all of the continents were closely connected. We will take a deeper look at Pangea in our chapter on the flood.

On Day Four, Elohim creates the sun, moon and stars. We read in Genesis 1:

¹⁴ And God said, Let there be lights in the firmament of the heaven to divide the day from the night; and let them be for signs, and for seasons, and for days, and years: ¹⁵ And let them be for lights in the firmament of the heaven to give light upon the earth: and it was so. ¹⁶ And God made two great lights; the greater light to rule the day, and the lesser light to rule the night: he made the stars also. ¹⁷ And God set them in the firmament of the heaven to give light upon the earth, ¹⁸ And to rule over the day and over the night, and to divide the light from the darkness: and God saw that it was good. ¹⁹ And the evening and the morning were the fourth day.

Elohim's creation is full of statistical mathematical probabilities that are impossible to have happened by chance. The sun and moon appear to us as exactly the same size in the sky. This is only possible because the moon is 1/400ᵗʰ the size of the sun and the sun is 400 times farther away from Earth than the moon. This distance is the only reason we're able to see total solar eclipses.

Evolutionists have said that this is a happy coincidence because millions of years ago the moon was closer. Once again, evolutions magic wand of time swoops in to save them from having to admit that they were created by an intelligent Being.

On Day Five, Elohim creates the birds and sea creatures. We read in Genesis 1:

> [20] And God said, Let the waters bring forth abundantly the moving creature that hath life, and fowl that may fly above the earth in the open firmament of heaven. [21] And God created great whales, and every living creature that moveth, which the waters brought forth abundantly, after their kind, and every winged fowl after his kind: and God saw that it was good. [22] And God blessed them, saying, Be fruitful, and multiply, and fill the waters in the seas, and let fowl multiply in the earth. [23] And the evening and the morning were the fifth day.

In our chapter on the animals, we will get into the massive antediluvian creatures that swam the seas. We'll also take a look at some of the terrifying pterosaurs that flew and may still fly our skies today.

On Day Six, Elohim creates land animals and humans. We read in Genesis 1:

> [24] And God said, Let the earth bring forth the living creature after his kind, cattle, and creeping thing, and beast of the earth after his kind: and it was so. [25] And God made the beast of the earth after his kind, and cattle after their kind, and every thing that creepeth upon the earth after his kind: and God saw that it was good. [26] And God said, Let us make man in our image, after our likeness:

and let them have dominion over the fish of the sea, and over the fowl of the air, and over the cattle, and over all the earth, and over every creeping thing that creepeth upon the earth. [27] So God created man in his own image, in the image of God created he him; male and female created he them. [28] And God blessed them, and God said unto them, Be fruitful, and multiply, and replenish the earth, and subdue it: and have dominion over the fish of the sea, and over the fowl of the air, and over every living thing that moveth upon the earth. [29] And God said, Behold, I have given you every herb bearing seed, which is upon the face of all the earth, and every tree, in the which is the fruit of a tree yielding seed; to you it shall be for meat. [30] And to every beast of the earth, and to every fowl of the air, and to every thing that creepeth upon the earth, wherein there is life, I have given every green herb for meat: and it was so. [31] And God saw every thing that he had made, and, behold, it was very good. And the evening and the morning were the sixth day.

The Creation of Adam by Michelangelo around 1512

The Hebrew word for man is **'āḏām**. Adam ends up being called by his race. Elohim created the Adams and blessed them Genesis 5:2, *"Male and female created he them; and blessed them, and called their name Adam, in the day when they were created."*

Notice in Genesis 1:30 that God gave green herbs to man and beast for food. This means that sharks and T. rexes ate vegetation at creation. There are three events in scripture that were likely responsible for the decline into carnivorism: The fall of creation, the corruption of the Nephilim and/or God's covenant with the Earth through Noah.

Genesis 3:17-18 tells us of the fall of creation, *"And unto Adam he said, Because thou hast hearkened unto the voice of thy wife, and hast eaten of the tree, of which I commanded thee, saying, Thou shalt not eat of it: cursed is the ground for thy sake; in sorrow shalt thou eat of it all the days of thy life; Thorns also and thistles shall it bring forth to thee; and thou shalt eat the herb of the field."*

At this point, creation itself is cursed. Now the plants grow thorns. If plants suddenly had the ability to grow sharp thorns, it's not a stretch that some animals developed sharp teeth.

Another possibility for carnivorism is the corruption of the Nephilim. Some speculate that the Nephilim mixed the DNA of the animals and that they are responsible for creating violent beasts. We read about the corruption in the book of Enoch, chapter 7:4-6, *"the giants turned against them and devoured mankind. And they began to sin against birds, and beasts, and reptiles, and fish, and to devour one another's flesh, and drink the blood. Then the earth laid accusation against the lawless ones."*

We also read about the corruption of the Nephilim in Genesis 6:11-12, *"The earth also was corrupt before God, and the earth was filled with violence. And God looked upon the earth, and, behold, it was corrupt; for all flesh had corrupted his way upon the earth."*

With the violence that the Watchers and their offspring the Nephilim brought to Elohim's creation, many of the animals were likely tainted. We will get into more of those details in our chapters on the Watchers and Nephilim.

Animals that had not become carnivores through the fall of creation or the corruption of the Nephilim, may have become so at God's postdiluvian covenant with the Earth. We read about some of the covenant in Genesis 9:

> [2] And the fear of you and the dread of you shall be upon every beast of the earth, and upon every fowl of the air, upon all that moveth upon the earth, and upon all the fishes of the sea; into your hand are they delivered. [3] Every moving thing that liveth shall be meat for you; even as the green herb have I given you all things. [4] But flesh with the life thereof, which is the blood thereof, shall ye not eat. [5] And surely your blood of your lives will I require; at the hand of every beast will I require it, and at the hand of man; at the hand of every man's brother will I require the life of man.

Before the flood, the animals were innocent. Their innocence likely made them easy prey for the Nephilim. After the flood, God changes the animal's mindset, and they avoid mankind.

The Power of God's Voice

God reveals Himself through creation. All anyone has to do is look with an innocent heart to see His existence. This is backed up in Romans 1:20-21, *"For the invisible things of him from the creation of the world are clearly seen, being understood by the things that are made, even his eternal power and Godhead; so that they are without excuse: Because that, when they knew God, they glorified him not as God, neither were thankful; but became vain in their imaginations, and their foolish heart was darkened."*

Nothing connects us to each other and the universe in the same way that sound does. Some scientists have theorized that our atoms, indeed everything in the universe, is made up of sound waves. How did Elohim create? Did He use His hands as Psalm 95 poetically says? No, Elohim created with His voice. He spoke it and it came to be. This idea is backed up in Hebrews 11:3, *"Through faith we understand that the worlds were framed by the word of God, so that things which are seen were not made of things which do appear."* And in Psalm 33:9, *"For he spake, and it was done; he commanded, and it stood fast."*

Because space is a vacuum, sound does not travel through it. This makes space the ideal vocal creation medium the same way you use a sink to keep water from splashing out while you wash.

The power of God's voice is a fascinating topic. Visualized in scripture, it likely made the mouth and face of the speaker appear distorted as in a pulse. Many times, in scripture, the voice of God is described as fire. In Jewish writings from the 12th century A.D., specifically the Shemot Rabba 5:9, we read a good description of the idea:

> On the occasion of the giving of the Torah, the Children of Israel not only heard the LORD's Voice but actually saw the sound waves as they emerged from the LORD's mouth. They visualized them as a fiery substance. Each commandment that left the LORD's mouth traveled around the entire camp and then came back to every Jew individually.

Psalm 29:7 reads, *"The voice of the Lord divideth the flames of fire."* In Acts 2 we read of another experience,

> [2] And suddenly there came a sound from heaven as of a rushing mighty wind, and it filled all the house where they were sitting. [3] And there appeared unto them cloven tongues like as of fire, and it sat upon each of them. [4] And they were all filled with the Holy Ghost, and began to speak with other tongues, as the Spirit gave them utterance....[6] Now when this was noised abroad, the multitude came together,

and were confounded, because that every man heard them speak in his own language.

Here in Acts, the power of God's voice is enough to undo the verbal confusion at Babel. Into how many nations did God separate the world at Babel? We read in Deuteronomy 32:8, *"When the Most High divided to the nations their inheritance, when he separated the sons of Adam, he set the bounds of the people according to the number of the children of Israel."*

God divides the world into 70 nations – the number of the children of Israel. How many disciples did Jesus send into the world with the ability to speak and every man hear in his own language? The answer is 70 in Luke 10:1, *After these things the Lord appointed other seventy also, and sent them two and two before his face into every city and place, whither he himself would come.*

Jesus reversed Babel. That's just bonus, so enjoy.

The Stars

Saint Augustine said, "Every visible thing in the world is put under the charge of an angel." Is there more to stars than just remote incandescent bodies, or suns, in our universe? We will see in our chapter on the Watchers that Elohim placed angels around creation to watch over it. There exists much in scripture and history that tell us that stars are associated with gods or angels.

In Judges 5:19-20, the stars fight against Sisera, *"The kings came and fought, then fought the kings of Canaan in Taanach by the waters of Megiddo; they took no gain of money. They fought from heaven; the stars in their courses fought against Sisera."*

Is this poetic language or are angels warring against the principalities of the air of Sisera? We will study about the angelic princes of the nations in the chapter on the postdiluvian Earth.

The **Enuma Elish** is a Sumerian creation story. It says that Marduk made the stars as stations for the great gods in Tablet 5: *He (Marduk) made the stations for the great gods; The stars, their images, as the stars of the Zodiac, he fixed.*

In Tablet 7 as the gods are praising Marduk, they again, refer to the stars as gods: *For the stars of heaven he upheld the paths, He shepherded all the gods like sheep!*

In the **Egyptian Book of the Heavenly Cow**, Ra creates gods to watch the Earth and they are the Milky Way:

> The Majesty of this god said, "Stay far away from them! humanity Lift me up! Look at me!" and so Nut became the sky Then the Majesty of this god was visible within her. She said, "If only you would provide me with a multitude to help me!" and so the Milky Way came into being.

As we are about to see, angels are mentioned as stars, stones of fire and burning mountains in scripture and in apocryphal literature. In Ezekiel 28, angels are referred to as stones of fire when Lucifer is cast down to the Earth:

> [13] Thou hast been in Eden the garden of God; every precious stone was thy covering, the sardius, topaz, and the diamond, the beryl, the onyx, and the jasper, the sapphire, the emerald, and the carbuncle, and gold: the workmanship of thy tabrets and of thy pipes was prepared in thee in the day that thou wast created. [14] Thou art the anointed cherub that covereth; and I have set thee so: thou wast upon the holy mountain of God; thou hast walked up and down in the midst of the stones of fire. [15] Thou wast perfect in thy ways from the day that thou wast created, till iniquity was found in thee. [16] By the multitude of thy merchandise they have filled the midst of thee with violence, and thou hast sinned: therefore I will cast thee as profane out of the mountain of God: and I will destroy thee, O covering cherub, from the midst of the stones of fire.

In the book of Enoch, the angels are referred to as stars and burning mountains. We also see the mountains of precious stones. We read in Enoch 18:

> [6.] And I proceeded and saw a place which burns day and night, where there are seven mountains of magnificent stones, three towards the east, and three towards the south...[13.] I saw there seven stars like great burning mountains, and to me, when I inquired regarding them,[14.] The angel said: 'This place is the end of heaven and earth: this has become a prison for the stars and the host of heaven. [15.] And the stars which roll over the fire are they which have transgressed the commandment of the Lord in the beginning of their rising, because they did not come forth at their appointed times.

The seven angels and the burning mountains here in Enoch are also seen in the book of Revelation chapter 8:6, 8, 10, *"[6]And the seven angels which had the seven trumpets prepared themselves to sound ...[8]And the second angel sounded, and as it were a great mountain burning with fire was cast into the sea: and the third part of the sea became blood ...[10]And the third angel sounded, and there fell a great star from heaven, burning as it were a lamp, and it fell upon the third part of the rivers, and upon the fountains of waters;"*

In Revelation 9:1, a star is a heavenly being, *"And the fifth angel sounded, and I saw a star fall from heaven unto the earth: and to him was given the key of the bottomless pit."*

Not only does God use the stars to measure the seasons, but He also uses them for signs. We read in Genesis 1:14, *"And God said, Let there be lights in the firmament of the heaven to divide the day from the night; and let them be for signs, and for seasons, and for days, and years"*

In Matthew 2:2 and 9, the star that leads the wise men to baby Jesus appears in the east, *"Saying, Where is he that is born King of the Jews? for we have seen his star in the east, and are come to worship him... When they had heard the king, they departed; and, lo, the star, which they saw in the east, went before them, till it came and stood over where the young child was."*

There was another time in history in which a star appeared. On July 4, 1054 A.D., a Jupiter-sized star appeared in the sky just below Orion. It stayed in the sky for almost two years until it departed on 17 April 1056. Astronomers believe this account because we have written records from both Chinese and Japanese astronomical texts. The texts say that the star was visible by day and pointed rays shot out from it on all sides.

If stars are for signs, then what of significance happened in 1054 A.D.? Twelve days after the star appeared, the Christian church had its first major split known as the Great Schism. The church was split into the Western Roman Catholic Church and the Eastern Orthodox Church. Was this schism the event foretold by the star? One can't be sure, but, like Nibiru, the idea of a visiting star is fascinating.

Angels are in the spiritual realm. Do they somehow abide in the stars that we see here in the physical? Do angels give the stars power to shine? As we studied in my last book, *Ancient Cities and the gods Who Built Them*, human skin can illuminate. In scripture, this happens when Moses spends time with God. In the book of Exodus, the Israelites ask Moses to cover his face because it's too bright. Similarly, do stars illuminate because of a spiritual being residing there? The theory is plausible.

Eden

The Garden of Eden is a fascinating place that gives us insight into what our purpose is and why we die. In Genesis 2:8 and 16, we read about the garden:

> [8] And the Lord God planted a garden eastward in Eden; and there he put the man whom he had formed. [9] And out of the ground made the Lord God to grow every tree that is pleasant to the sight, and good for food; the tree of life

also in the midst of the garden, and the tree of knowledge of good and evil...[16] And the Lord God commanded the man, saying, Of every tree of the garden thou mayest freely eat: [17] But of the tree of the knowledge of good and evil, thou shalt not eat of it: for in the day that thou eatest thereof thou shalt surely die.

Why did God place a tree of which the Adams were not allowed to partake? Why do bad things happen to good people? I believe it's because God gave humankind free will. We could be autonomous robots and do exactly what God has planned, but He, in His infinite wisdom, gave us a choice.

Adam and Eve are driven out of the Garden of Eden

After Adam and Eve ate of the Tree of the Knowledge of Good and Evil, why were they kicked out of the garden? Elohim knew that mankind would eventually eat from the Tree of Life. We read in Genesis 3:22-24, *"And the Lord God said, Behold, the man is become as one of us, to know good and evil: and now, lest he put forth his hand, and take also of the tree of life, and eat, and live for ever: Therefore the Lord God sent him forth from the garden of Eden, to till the ground from whence he was taken. So* he drove out the man; and he placed at the east of the garden of Eden Cherubims, and a flaming sword which turned every way, to keep the way of the tree of life."

Notice in verse 22 that Elohim speaks in the plural, *"the man is become as one of us."* I don't believe that this is referring to the Holy Trinity, but to the heavenly council. We will explore more about the Psalm 82 council of Elohim in the chapter on the Watchers.

Ponce de Leon and his supposed search for the fountain of youth may not be as farfetched as we are told. If you eat from the Tree of Life, you will become immortal. If you could find the Tree of Life, would you even want to, considering it's guarded by a flaming sword?

Interestingly, scripture says that man was placed in the garden after God created both man and Eden. We read about this in Genesis 2:8, *"And the Lord God planted a garden eastward in Eden; and there he put the man whom he had formed."*

Does "eastward in Eden" mean that there was something in the west or is it just written from the perspective of the physical location of the person recording Genesis?

I speculate that Eden was somewhere between the spiritual and physical realms – maybe in both simultaneously. There's a lot of scriptural evidence to back up this theory. As we already read in Ezekiel 28:13-14, Eden was likely on the Holy Mountain of God, *"Thou hast been in Eden the garden of God; every precious stone was thy covering, the sardius, topaz, and the diamond, the beryl, the onyx, and the jasper, the sapphire, the emerald, and the carbuncle, and gold: the workmanship of thy tabrets and of thy pipes was prepared in thee in the day that thou wast created. Thou art the anointed cherub that covereth; and I have set thee so: thou wast upon the holy mountain of God; thou hast walked up and down in the midst of the stones of fire."*

Psalm 2:6 calls this holy mountain Zion, *"Yet have I set my king upon my holy hill of Zion."*

It sounds like the millennial kingdom will be a redo of creation before the flood and before the fall. In the future millennial kingdom, we see God's Holy Mountain in Isaiah 65:25, *"The wolf and the lamb shall feed together, and the lion shall eat straw like the bullock: and dust shall be the serpent's meat. They shall not hurt nor destroy in all my holy mountain, saith the Lord."*

There were also angels with bad intentions in the garden according to both the Bible and the book of Enoch. The Garden of Eden was on the Holy Mountain of God and is where Lucifer is cast down in Ezekiel 28:

> [14] Thou art the anointed cherub that covereth; and I have set thee so: thou wast upon the holy mountain of God; thou hast walked up and down in the midst of the stones of fire. [15] Thou wast perfect in thy ways from the day that thou wast created, till iniquity was found in thee. [16] By the multitude of thy merchandise they have filled the midst of thee with violence, and thou hast sinned: therefore I will cast thee as profane out of the mountain of God: and I will destroy thee, O covering cherub, from the midst of the stones of fire. [17] Thine heart was lifted up because of thy beauty, thou hast corrupted thy wisdom by reason of thy brightness: I will cast thee to the ground, I will lay thee before kings, that they may behold thee.

Scripture does not tell us much of who the serpent was who tempted Eve. We will look at the origins of the conflict between spirits and mankind in the chapter on the Adams. The book of Enoch tells us that the serpent was a fallen angel named Gadreel in chapter 69:6, *"And the name of the third is Gadreel; this is the one that showed all the deadly blows to the sons of men. And he led astray Eve. And he showed the weapons of death to the children of men, the shield and the breastplate, and* the sword for slaughter, and all the weapons of death to the sons of men."

Revelation chapters 2, 21 and 22 give us more evidence that Eden was closer to the spiritual realm than we are today. The great and high mountain of God and the Tree of Life are in Paradise. We read in Revelation 2:7, *"He that hath an ear, let him hear what the Spirit saith unto the churches; To him that overcometh will I give to eat of the tree of life, which is in the midst of the paradise of God."*

We read in Revelation 21:10, *"And he carried me away in the spirit to a great and high mountain, and shewed me that great city, the holy Jerusalem, descending out of heaven from God,"*

We read in Revelation 22: 1-2, *"And he shewed me a pure river of water of life, clear as crystal, proceeding out of the throne of God and of the Lamb. In the midst of the street of it, and on either side of the river, was there the tree of life, which bare twelve manner of fruits, and yielded her fruit every month: and the leaves of the tree were for the healing of the nations."*

These scriptures show evidence that the Garden of Eden may have been closer to the spiritual realm than our Earth is today.

Where are the garden and the Tree of Life now? Is the garden in the Bermuda Triangle? Is it under the ice in Antarctica? Is it at the North Pole? I speculate that it was taken from the physical realm altogether and that it will reappear sometime in or after the millennium.

There is some evidence suggesting the Garden of Eden was pushed to the North Pole after the great flood. Many times, in scripture, Heaven is said to have been located in the north.

Leviticus 1:11, *"And he shall kill it on the side of the altar northward before the Lord: and the priests, Aaron's sons, shall sprinkle his blood round about upon the altar."*

Psalm 48:2, *"Beautiful for situation, the joy of the whole earth, is mount Zion, on the sides of the north, the city of the great King."*

In Psalms 75:6, we read that salvation comes from the north, *"For promotion cometh neither from the east, nor from the west, nor from the south."*

Job 26:7, *"He stretcheth out the north over the empty place, and hangeth the earth upon nothing."*

Isaiah 14:13, *"For thou hast said in thine heart, I will ascend into heaven, I will exalt my throne above the stars of God: I will sit also upon the mount of the congregation, in the sides of the north:"*

As we can see, scripture indicates that Heaven is in the direction of the north. God's abode is in the spiritual dimension and not our physical one. Is that location to our north?

If you ask a child to point to Heaven, they will likely point upward. The direction, up, as it relates to our cosmos is constantly changing. If, however one was to point to the North, the location would be constant as there is a point in our cosmos to which the Earth's axis points. The north star is not that point but it's close and is why navigators use it as a constant.

There is another phenomenon to our north – the aurora borealis. Are the spectacular northern lights an underside reflection of God's throne room with its emerald rainbow and sapphire floor? Southern lights exist which are called the aurora australis but they're not nearly as majestic. Revelation 4:3,6 tells us of the emerald rainbow that surrounds God's throne: *"³And he that sat was to look upon like a jasper and a sardine stone: and there was a rainbow round about the throne, in sight like unto an emerald ...⁶And before the throne there was a sea of glass like unto crystal: and in the midst of the throne, and round about the throne, were four beasts full of eyes before and behind."*

Gerhard Mercator was a Flemish cartographer who is best known for the Mercator projection, the method of taking the curved lines of the Earth and transforming them into straight ones that can be used by sailors on a flat map. Mercator charted the first full map of the Arctic in 1569. The map depicts the Artic as a mountain out of which four rivers flow surrounded by four large islands. We read about the four rivers that flowed out of Eden in Genesis 2:

> [10] And a river went out of Eden to water the garden; and from thence it was parted, and became into four heads. [11] The name of the first is Pison: that is it which compasseth the whole land of Havilah, where there is gold; [12] And the gold of that land is good: there is bdellium and the onyx stone. [13] And the name of the second river is Gihon: the same is it that compasseth the whole land of Ethiopia. [14] And the name of the third river is Hiddekel: that is it which goeth toward the east of Assyria. And the fourth river is Euphrates.

Mercator explained his map of the Artic in a 1577 letter to John Dee, an English mathematician and astrologer:

> In the midst of the four countries is a Whirl-pool... into which there empty these four indrawing Seas which divide the North. And the water rushes round and descends into the earth just as if one were pouring it through a filter funnel. It is four degrees wide on every side of the Pole, that is to say eight degrees altogether. Except that right under the Pole there lies a bare rock in the midst of the Sea. Its circumference is almost 33 French miles, and it is all of magnetic stone...

This magnetic rock is also known as the Rupes Nigra ("Black Rock"). It purportedly explained why all compasses pointed to the North Pole.

A 1569 map of the Arctic by Gerhard Mercator

For years, Google withheld images of the Arctic on Google Earth. Now they show nothing but water and ice at the North Pole. Was Mercator wrong with his cartography of the Arctic or are we being deceived yet again by powers that will do anything to destroy evidence that could confirm God's Word?

Is the Garden of Eden located at the North Pole? It's one of our mysteries that's not impossible to figure out. One day, you or I may take an expedition to see for ourselves. Of course, would you even want to find the garden, considering there are cherubim's (winged angels) guarding the gates and a flaming sword "keeping the way" of the Tree of Life?

Secular Creation Stories

One mythological creation story that has similar events as creation in scripture is that of the Anunnaki and planet Nibiru. This story comes from Sumerian records kept by Ashurbanipal.

According to author Zecharia Sitchin, the 12th planet known as Nibiru, doesn't have a circular orbit around the sun but a long elliptical one. Once every 3,600 years, Nibiru comes close enough to Earth for the alien inhabitants, the Anunnaki, to visit our planet for a time.

These texts say that the Anunnaki needed beings to mine gold for them among other tasks. The Anunnaki mixed their own DNA with that of Earth's hominids to create a semi-intelligent workforce called Igigi.

Three thousand six hundred years later, when Nibiru was close enough, the Anunnaki visited again. They perfected their earlier creation, gave them the ability to reproduce, gave them intelligence and called them Adamu.

We don't see the giant fiery red planet Nibiru in our skies today. If it is a real planet, then maybe it's not close enough yet to see. Scientists do agree that there is a massive unseen body in our solar system and its gravitational pull is what's causing the Sun to tilt 6 degrees.

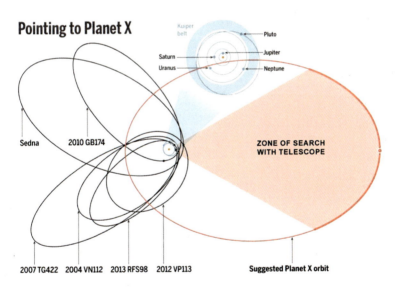

The projected orbit of a massive planet (maybe Nibiru) tilting the sun 6 degrees

The **Enuma Elish** is the Babylonian story of creation written sometime before 1600 B.C. This story details the creation events, how Marduk ruled the cosmos and how the city of Babylon came to be. According to the **Enuma Elish**, in the beginning, there existed only water:

When in the height heaven was not named,
And the earth beneath did not yet bear a name,
And the primeval Apsu, who begat them,
And chaos, Tiamut, the mother of them both
Their waters were mingled together,
And no field was formed, no marsh was to be seen;
When of the gods none had been called into being

The gods warred among themselves. Marduk eventually kills Tiamat and uses her body to create the Earth. Out of clay, he created Lullu as the first man according to Sumerian texts:

In the clay, god and man
Shall be bound,
To a unity brought together;
So that to the end of days
The Flesh and the Soul
Which in a god have ripened –
That soul in a blood-kinship be bound

There's quite a parallel to Elohim's creation of Adam and breathing the breath of life into him. This creation story also parallels with the Bible in that before everything else, there was water and that the gods warred in Heaven.

The **Atrahasis** is the Akkadian/Babylonian epic written around 1700 B.C. In the story, Enki, the god of wisdom, created mankind from clay to serve the gods by toiling on the Earth.

In the **Epic of Gilgamesh**, Enkidu was the first man. He was formed from clay as well. We read on Tablet 1:, *"Aruru Finger'd some clay, on the desert she moulded 4 (it): (thus) on the desert. Enkidu made she, a warrior, (as he were) born (and) begotten, (Yea), of Ninurta 5 the double..."*

What we see is a common theme among ancient creation stories: Heaven and Earth were created and separated by water. Mankind was created to serve the gods. Why are all of these ancient stories so similar? It's as if everyone on the planet is descended from one family with the same story. Maybe there's more to our mythologies than just myth.

THE FIRMAMENT

According to the Bible, as well as other ancient sources, life before the flood of Noah was much different than it is today. Antediluvian humans lived eight times as long as we do. Lizards as well, as other animals, grew to enormous sizes. The skies didn't give rain. Giants lived among humans. Today, many of the antediluvian animals are extinct. Mankind has much shorter lifespans. Most have never seen a giant nor heard stories from those that have.

It's obvious that something drastically changed at the flood of Noah that altered Elohim's creation forever. A major change was the breaking and collapse of the ice canopy above the firmament – the one thing God did not pronounce good at creation. It's a reasonable theory that the canopy was a layer of ice that encapsulated the Earth before the flood of Noah. Many ancient cultures called it a dome.

The ice canopy that likely covered the antediluvian Earth

Notice in Isaiah 40:22, that God stretches out the heavens as a curtain or tent, *"It is he that sitteth upon the circle of the earth, and the inhabitants thereof are as grasshoppers; that stretcheth out the heavens as a curtain, and spreadeth them out as a tent to dwell in"*

As a side note, evolutionists scoff at God and say that Old Testament authors believed that the Earth was flat. They are wrong in their attempts to poke fun at their Creator. Isaiah 40 clearly states that the Earth is round.

I speculate that in the days after creation, the water above froze and turned into a layer of ice. The book of Job contrasts the canopy first in its liquid form, and then in its solid form. In Job 26:11, we read, *"The pillars of heaven tremble and are astonished at his reproof."*

The Hebrew word used for "tremble" is *rûp̄.* The next best translation of *rûp̄* is to quiver or flutter. The pillars of Heaven trembling I believe refers to the coagulating of the water as it freezes. In Job 37:18, we read, *"Hast thou with him spread out the sky, which is strong, and as a molten looking glass?"*

Molten here comes from the Hebrew word *mu·ṣaq* meaning to cast metal. The **Genesis Rabbah** (Bereishit 4) describes the transformation of the water from a liquid to a solid:

> Sa-mayim ("carry water"), i.e., it holds up water. The heavens were like milk in a dish; until a drip of rennet is put in it, the milk sloshes, but once a drop of milk is put in it, the milk congeals and becomes solid. This is the meaning of (Job 26:11): 'The pillars of the heavens were shaky.' A coagulant was placed in it, "and it was evening and it was morning, day two.

The idea of water encapsulating the Earth is also referred to as the Canopy Theory. The Hebrew word used for firmament in Genesis 1 is *rā·qî·a'* meaning an extended surface or expanse. *Rā·qî·a'* comes from the root word raqa meaning to stamp or beat out like one would hammer a sheet of metal.

In Exodus 39:3, the Hebrew uses *way·raq·qǝ·'ū* (And they did beat) when describing how the Israelites made the tabernacle contents, *"And they did beat the gold into thin plates, and cut it into wires, to work it in the blue, and in the purple, and in the scarlet, and in the fine linen, with cunning work."*

At the flood, the breaking up of the fountains of the deep likely shot rocks into the ice canopy so violently that the canopy broke all around the world. We'll get into the separation of Pangea in a later chapter about the flood, so I'll digress.

The ice canopy would have created a greenhouse effect on Earth. The greenhouse effect is the process of: a barrier letting light in, the ground absorbing the light energy, the ground re-emitting the light energy and the original barrier not letting the re-emitted energy out.

The Earth's greenhouse effect would be total – from north pole to south pole. The worldwide temperature would range between 72 and 78 degrees Fahrenheit. In fact, scientists agree that specimens at the bottom layers of the fossil record exhibit warm tropical temperatures.

In 1961, dinosaur bones were discovered in the Alaskan tundra by Shell Oil Co. This represented the first find of dinosaurs at high and currently cold altitudes.

The Earth would have had no weather patterns but perfect weather year-round. Currently the Earth is tilted on its axis by about 23 degrees. If the Earth were perfectly vertical on its axis, it would be what we know as springtime year-round.

Some will say that you couldn't see the stars through an ice canopy. I believe that not only could you see the stars through the ice canopy, but the canopy would be a photo amplifier making the stars easier to see than we can today.

If you do much rifle shooting on a hot day as I did during my time serving as a Marine Corps Scout Sniper, you'll notice heat waves coming off the ground. The waves are actually a great way to tell wind direction when you're shooting at a distance.

The heat waves distort and blur what you're looking at behind them. Our atmosphere does the same thing when we look up at the stars. This causes atmospheric twinkle which is why stars twinkle in our vision.

The ice canopy would have compressed our atmosphere so that it had less atmospheric distortion. This would make the stars more visible than they are today. In Genesis 1:16-17, God sets the stars in the firmament, *"And God made two great lights; the greater light to rule the day, and the lesser light to rule the night: he made the stars also. And God set them in the firmament of the heaven to give light upon the earth"*

The Hebrew word for "set" is **nathan** which means to give or to add. God did not put the stars physically in the firmament, but from man's perspective (who the Bible was written for) the firmament would absolutely show the stars in a different and more majestic way.

The sun's effects have been diluted, thus enabling humans to live extremely long lives and lowering cancer rates drastically. These conditions would almost eliminate genetic mutations caused by UV rays. We explore this topic further in the chapter on the Adams.

Amber is fossilized tree resin in which we find many things from the antediluvian world. Plenty of insects are found entombed inside of the amber. Today, these insects look just like they did "millions" of years ago. Isn't it amazing that they haven't evolved?

We also find air bubbles trapped inside of amber. The oxygen content of those air bubbles is 32 percent, 11 percent higher than the oxygen we have in our atmosphere today. With 32 percent oxygen, compared to today's 21 percent, a fit human would be able to run for hours without getting fatigued. Dinosaurs with their tiny lungs and nostrils would have been able to breathe!

An ensign wasp entombed in amber. Credit: Oregon State University, Wikimedia

Atmospheric pressure would be 1.6 times what it is today. With more pressure, wounds heal much more quickly. We artificially do this today through a process called Hyperbaric Oxygen Therapy. It's a medical procedure that pressurizes oxygen on a wound or in the body in order to heal it more quickly. With Hyperbaric Oxygen Therapy, doctors treat multiple sclerosis, arthritis, vascular necrosis, cirrhosis, leprosy, diabetic gangrene, Burger's disease, advanced liver disease, decompression sickness and burns.

Humans are not the only ones to benefit from a more pressurized atmosphere. In hyperbaric conditions, plants grow larger and healthier. In the 1980s, Dr. Kei Mori conducted experiments with plants at Keio University in Tokyo. He built special chambers in which to grow plants by blocking out IR and UV radiation. To do this, Dr. Mori wired fiber optic cables to bring light to the plants in a process called Himawari Sunlighting. He pressurized carbon dioxide into the stems of the tomato plants with a rubber sock.

In one of his experiments, he grew a tomato plant to over 30 feet high. This tree was displayed to the public at the Japan Expo of 1985. During the expo, the "tomato tree" produced more than 13,000 ripe tomatoes in a six-month span!

Dr. Dan Carson conducted misting and birdsong experiments with plants. This "Sonic Bloom" process was birthed in the 1970s. Sonic simulation (birdsong) causes the stomata (breathing holes) under leaves to open wider, allowing more carbon dioxide and nutrients to enter plants.

During Dr. Carson's experiments, he grew a purple passion plant, which normally grows to 18 inches, to a world record 1,300 feet in height. He grew 10-inch potatoes and cantaloupes. Apples using the process had a shelf life of five months instead of the normal 30 days. A 2003 world record pumpkin weighing 1,458 pounds was grown using this technique.

As a quick and interesting side note, fermentation of alcohol is slower in higher pressure climates. Genesis 9:20-21 tells us that Noah became drunk with wine after the flood, *"And Noah began to be an husbandman, and he planted a vineyard: And he drank of the wine, and was drunken; and he was uncovered within his tent."*

Did Noah not expect his grape wine to be as potent as it was because he was used to the antediluvian conditions?

As we can see through modern experimentation, plants would have grown to be massive in the antediluvian world. The nutritional content of these plants would have been extremely high due to the conditions as well as more pure genetics. As we will see in future chapters, animals and man benefitted greatly from this super food source.

Biblical Evidence

It was John Calvin, the founder of Calvinism, who popularized the biblical account of the firmament as only being the clouds that we see today. He did so to encourage 16th-century Christians to accept the findings of science. However, there is much evidence in scripture for the firmament being a solid dome. We read in Genesis 2:5-6, *"And every plant of the field before it was in the earth, and every herb of the field before it grew: for the Lord God had not caused it to rain upon the earth, and there was not a man to till the ground. But there went up a mist from the earth, and watered the whole face of the ground."*

Here in Genesis 2, God had not caused it to rain on the Earth, so He sent a mist up from the ground to water the flora. Can we see evidence of this today? According to a Discover Magazine article, "The Strange Forests that Drink – and Eat – Fog," flora in tropical rainforests today get 75 percent of their water from moisture in the air. The fact that the Earth's flora didn't get water from rain, is evidence that clouds were not a part of the antediluvian world.

A mist waters this tropical rainforest in Asia

We ask a lot of questions within the bounds of God's word. Mysteries are mysteries because they don't have an easily ready answer. In our attempts to find missing puzzle pieces, we often end with a question instead of an answer.

What connection does the mist from the ground have to the fountains of the great deep that broke up at the time of the flood? Was the subterranean makeup different somehow. How was there enough mist from the ground to water all of creation?

Also note that four rivers flowed from the Garden of Eden. It would have to have been a massive water flow coming from the fountains of the great deep that supplied those four rivers from one origin.

Hebrews 11:7 tells us that Noah built the ark despite not having seen some things before, *"By faith Noah, being warned of God of things not seen as yet, moved with fear, prepared an ark to the saving of his house; by the which he condemned the world, and became heir of the righteousness which is by faith."*

As a rabbit trail, notice here that Noah built the ark to save his house. If you've seen the 2014 Hollywood adaptation of the great flood in the movie **Noah**, you'll know that they grossly miss the mark of scripture. In the film, Noah thinks it's his calling to only allow the animals to survive the flood but that mankind will be blotted out.

While the film falls short of scripture, there were a few things that I thought interesting and plausible interpretations of the antediluvian world. The films' strange creatures were absolutely something that I think the film got right. Not only do up to 2,000 of all animal species go extinct every year, but the Nephilim, I believe, were mixing animal DNA and creating strange hybrids or chimeras. We'll get more into that in the chapter, The Nephilim.

In the film, the portrayal of mankind as industrial and Earth-consuming was a good interpretation. I do believe that antediluvian mankind had reached levels of technology that almost no one gives proper credence to. We'll get into some of that technology, and ancient stories about that technology, in the chapter on the Adams.

Also in the **Noah** film, Tubal-cain, and mankind in general, consumed the animals – which was forbidden by God in the antediluvian world. In one scene, as the animals are traveling to the ark, the sons of Cain kill some of the animals to be butchered for meat, ensuring the species was never seen again after the flood.

Back to the firmament. There is more evidence for a canopy in scripture. In Genesis 1:20, we read, *"And God said, Let the waters bring forth abundantly the moving creature that hath life, and fowl that may fly above the earth in the open firmament of heaven."*

Where the King James reads, "...in the open firmament...", the Hebrew translation reads *'al- pǝ·nê rǝ·qî·a'* or "across the face of the firmament." *Pǝ·nê* is also used in Genesis 1 verse 2 where God hovers over the face of the deep, *"And the earth was without form, and void; and darkness was upon the face of the deep. And the Spirit of God moved upon the face of the waters."*

Contrast that to Deuteronomy 47:18 where the fish swim in the water, *"The likeness of any thing that creepeth on the ground, the likeness of any fish that is in the waters beneath the earth"*

Psalm 148:4 tells us that there are waters above the heavens and that the raqia supported them, *"Praise him, ye heavens of heavens, and ye waters that be above the heavens."*

In Psalm 150:150, the raqia is God's place of power, *"Praise ye the Lord. Praise God in his sanctuary: praise him in the firmament of his power."*

Scripture tells us about the awesome visage of the canopy above the firmament. An ice layer would reflect the sun, moon and stars for the full 24-hour rotation of the Earth. It would never be completely dark at night. We read about visions of the firmament in Ezekiel 1:22, *"And the likeness of the firmament upon the heads of the living creature was as the colour of the terrible crystal, stretched forth over their heads above."*

The Hebrew word for terrible here is **han·nō·w·rā** which means a few things but "perfect" and "stand in awe of" are two of the translations. A layer of ice with whatever reflections from below or above would appear as crystal of different colors at different times of the day. We continue in Ezekiel 1:26-27, *"And above the firmament that was over their heads was the likeness of a throne, as the appearance of a sapphire stone... And I saw* *as the colour of amber, as the appearance of fire round about within it, from the appearance of his loins even upward, and from the appearance of his loins even downward, I saw as it were the appearance of fire, and it had brightness round about."*

Some scholars will say that the atmosphere had a pink hue due to the energized hydrogen in the air. A thick ice canopy would have given off a blue hue as ice absorbs red light. You'll notice that glaciers are a rich blue color because the red light doesn't penetrate deeply. I speculate that the ice canopy was only a few inches thick which would not stop much of the red light from passing through. The antediluvian sky would have looked fantastical.

Since 1640, biologists have known that plants grow bigger under pink light. The largest known fossil forest was discovered in a coal mine in Springfield, Illinois. It is at least 50 times larger than the previously largest discovered fossilized forest. Scale trees have been found in the mine with 6-foot diameters. Most of the planet likely looked more like the giant redwood forest we see today in California.

Fossilized Lepidodendrons (scale trees). Credit: OAnick, Wikimedia | Credit: Ghedoghedo, Wikimedia

Ancient Accounts

It's obvious that there is more truth to our ancient mythologies than most want to believe. Many of our mythologies and ancient cultures refer to the sky as having been a solid dome.

It's important to note that the firmament was likely destroyed at the great flood, and the cultures who write about it were repeating stories from their ancestors – the eight survivors that walked off of the ark. The cultures may have pulled the descriptions of the firmament from manuscripts of the ancient world that were taken onto the ark. It is possible that some of the firmament was left after the flood and slowly melted and fell apart in the years after the flood. There were other antediluvian effects that lingered and slowly dissipated over generations such as the age of man.

The Rig Veda is an ancient Indian collection of hymns that record the sky as a great dome. The ancient Hindus believed that "the earth was spread upon the cosmic waters" and that they "surrounded heaven and earth, separating the dwelling-place of men and gods…" *The Rig Veda* is believed to have been written as long ago as 1900 BC. That timeframe is around the biblical time of Joseph and 1,400 years after what I propose to be the year of the great flood (3,298 B.C.).

Rig Veda Mandal 9 Sukt 94

> Mantra 2 "Opening out the abode of the ambrosia (the firmament – the home of the waters) on both sides (he passes between); for him, the omniscient, the worlds expand. Gratifying laudations eager for the sacrifice, call upon Indu, like kine (lowing) towards their stall."

In this and other verses of *The Rig Veda*, the gods dwell in the great dome. It is the great ocean of happiness where nothing perishes.

The Sumerians wrote that the world was enclosed in a box of

imperishable tin. The Sumerian gods were known as Anunnaki meaning the "offspring of the lord." Anu was a Sumerian god who was the "Lord of the Firmament." Anu appears in the Epic of Gilgamesh and sends the Bull of Heaven to kill Gilgamesh. Anu and Igigi stand on stone floors in the Upper and Middle Heavens. This is synonymous with God standing on a blue sapphire brick heavenly floor in Exodus 24:9-10, *"Then went up Moses, and Aaron, Nadab, and Abihu, and seventy of the elders of Israel: And they saw the God of Israel: and there was under his feet as it were a paved work of a sapphire stone, and as it were the body of heaven in his clearness."*

We also read about the sapphire stone in Ezekiel 10:10, *"Then I looked, and, behold, in the firmament that was above the head of the cherubims there appeared over them as it were a sapphire stone, as the appearance of the likeness of a throne."*

The Babylonian Epic of Creation, the Enuma Elish contains explicit statements that the heavens are made of water. In Tablet IV, Marduk builds the heavens out of the watery corpse of Tiamat, a primordial goddess of the sea, *"137 He split her into two like a dried fish: 138 One half of her he set up and stretched out as the heavens. 139 He stretched the skin and appointed a watch 140 With the instruction not to let her waters escape."*

3 Baruch (the Greek Apocalypse of Baruch) was written between the 1st and 3rd centuries A.D. In 3rd Baruch 3, it tells of men, while building the tower of Babel, attempting to auger through the firmament:

> 6 And the Lord appeared to them and confused their speech, when they 7 had built the tower to the height of four hundred and sixty-three cubits. And they took a gimlet, and sought to pierce the heaven, saying, Let us see (whether) the heaven is made of clay, or of 8 brass, or of iron. When God saw this He did not permit them, but smote them with blindness and confusion of speech, and rendered them as thou seest.

In the Babylonian Talmud, Sanhedrin 109, the Rabbis tell us what the men of Babel said, *"What sin did they perform? Their sin is not explicitly delineated in the Torah. The school of Rabbi Sheila say that the builders of the Tower of Babel said: We will build a tower and ascend to heaven, and we will strike it with axes so that its waters will flow."*

Jubilees 2:4 from the Dead Sea Scrolls reads, *"And on the second day He created the firmament in the midst of the waters, and the waters were divided on that day – half of them went up above and half of them went down below the firmament (that was) in the midst over the face of the whole earth. And this was the only work (God) created on the second day."*

2 Enoch 3 reads, *"And it came about, when I had spoken to my sons, those med called me. And they took me up onto their wings, and carried me up to the first heaven, and placed me on the clouds. And, behold, they were moving. And there I perceived the air higher up, and higher still I saw the ether. And they placed me on the first heaven. And they showed me a vast ocean, much bigger than the earthly ocean."*

The Testament of Levi, from the Dead Sea Scrolls, in chapter 2:6-7 reads, *"And behold the heavens were opened, and an angel of God said to me, Levi, enter. And I entered from the first heaven, and I saw there a great sea hanging."*

Ancient Egyptian tomb writings tell of a watery sky with words such as "The watery abyss of the sky," "The Celestial Waters," "The Nile in heaven," and "The pool of the firmament." Nut, pronounced newt, was the Egyptian goddess of the firmament. They believed that by day the sun sailed across the surface of the sky-ocean named Canopus (likely where we get the word canopy from as the Greeks got **konopeion** from the Egyptians). The Egyptians, Hittites, Finns and Tibetans all described a domed sky made of iron.

In a native American story, Coyote, a young Indian boy climbs into the sky with an arrow ladder. He made the ladder by shooting arrows into the dome:

> Next evening Coyote saw they were all looking up in the sky at something. He asked the next oldest wolf what they were looking at, but he wouldn't say. It went on like this for three or four nights. No one wanted to tell Coyote what they were looking at because they thought he would want to interfere. One night Coyote asked the youngest wolf brother to tell him, and the youngest wolf said to the other wolves, "Let's tell Coyote what we see up there. He won't do anything."
>
> So they told him. "We see two animals up there. Way up there, where "we cannot get to them.
>
> "Let's go up and see them," said Coyote.
>
> "Well, how can we do that?"
>
> "Oh, I can do that easy," said Coyote. "I can show you how to get up there without any trouble at all."
>
> Coyote gathered a great number of arrows and then began shooting them into the sky. The first arrow stuck in the sky and the second arrow stuck in the first. Each arrow stuck in the end of the one before it like that until there was a ladder reaching down to the earth.
>
> "We can climb up now," said Coyote. The oldest wolf took his dog with him, and then the other four wolf brothers came, and then Coyote. They climbed all day and into the night. All the next day they climbed. For many days and nights they climbed, until finally they reached the sky. They stood in the sky and looked over at the two animals the wolves had seen from down below. They were two grizzly bears.

In Greek mythology, heaven was an ocean. It was a golden age where men lived long lives. After its destruction, the new god Zeus brought rain, hail, and snow. Helios, the sun god came into being at the same time.

The belief that the sky was a solid dome permeates our ancient world history. Though these stories may not be wholly true, I do believe that they all have some truth. I believe that one of those truths is that the sky was solid before the flood of Noah.

Whatever triggered the collapse of the firmament, it would have been an awesome thing to behold. We will explore the collapse of the firmament in more detail in the chapter on the flood.

Arguments Against

Some argue that the canopy theory is not possible because a layer of water vapor in our atmosphere would raise temperatures drastically below the canopy, thus cooking everything on Earth. I don't believe that this argument effectively dismantles the canopy theory.

Albedo is the measurement of how much of the Sun's light is reflected from Earth. Today, our planet reflects an average of 31 percent of the Sun's rays. Water reflects the least (7 percent), and snow reflects the most (80 percent). This process is more commonly known by black T-shirts absorbing the most heat while white T-shirts reflect the most. If the Albedo was raised to 70-plus percent, the ice canopy wouldn't cook the planet. The question becomes how was the Albedo raised.

I don't think it needs to be raised because this argument assumes that the canopy was composed of vapor instead of ice. When vapor turns into water, it releases heat, thus cooking the Earth. When ice turns to water, it absorbs heat. Therefore, this argument doesn't "hold water" against a canopy of ice.

Another argument against the canopy theory is the idea that pre-flood trees show signs of seasonality – that the trees endured the different seasons. Today we have discovered dozens of polystrate fossils.

Polystrate fossils are tree fossils that extend through more than one geological stratum. Trees that stand through multiple layers of stratum are an argument against evolution. Trees don't just stand upright for billions of years while layers of strata build up around them, eventually burying them.

There are plenty of localized catastrophes that can produce polystrate fossils. We cannot guarantee that these season-weathered trees are antediluvian and therefore cannot say with certainty that they disprove the canopy theory. These theories don't conflict. They both point to a young Earth.

Polystrate fossils extending through layers of strata. Credit: Michael Rygel, Wikimedia

We, of course, don't see the solid firmament today. Stick with me, because in our last chapter on the postdiluvian Earth, I speculate with great evidence that the firmament will return.

CHAPTER 3

THE ADAMS

Mankind is capable of both great and terrible things. We can show great compassion for others in a time of need. We can show great evil in a time of selfishness or pride.

Morals are not fluid. They are absolute. God tells us in His Word what is right and what is wrong. God never changes. He never has and never will. What He says is that sin will never change. If your "Jesus" is OK with any sin, then you are worshiping an idol – no matter how many times you call it Jesus. It all comes down to your priorities. Do you want to honor God or yourself?

The 2023 Grammy awards took place with an ode to Satan. Prominent musicians who claim to be Christians partook. They think they can fit their version of Christianity into their perfect world where they are accepted because they're one of the "cool" Christians. Like sheep to the slaughter, their paths lead to death.

Genesis 5:2, *"Male and female created he them; and blessed them, and called their name Adam, in the day when they were created."*

There was a time when man was innocent – a time before he knew what good and evil were. Once Eve ate the fruit, her eyes were opened and she understood what nakedness was. A dog doesn't know good from evil; it knows reward from consequence. Because of Adam and Eve's sin, we see, and therefore, are accountable for our sin.

I don't think Adam accidentally sinned by eating the fruit from the Tree of the Knowledge of Good and Evil. Before he took a bite of the apple, he knew exactly what he was doing. As he thought about it, he slumped his shoulders and made a conscious decision to disobey God – maybe out of loyalty to his wife, maybe out of a desire to be wise.

Angels had their part to play in our fall, too. We can't know for sure if the Adams would eventually eat from the tree on their own, but we do know that an angel tempted Eve. A war between angels and Adams had come.

In the chapters on the Watchers, we will study the corruption of the human bloodline. In the chapter on the Nephilim, we will look at the destructive war that the Nephilim wage against humanity. Here in Genesis 3, I believe, is where it all starts:

> [4] And the Lord God said unto the serpent, Because thou hast done this, thou art cursed above all cattle, and above every beast of the field; upon thy belly shalt thou go, and dust shalt thou eat all the days of thy life:...[15] And I will put enmity between thee and the woman, and between thy seed and her seed; it shall bruise thy head, and thou shalt bruise his heel.

In Hebrew, enmity is hatred. This is the beginning of using the bloodline as the point of contention between the fallen angels, the giants, the evil spirits and mankind. No. The giants and evil spirits don't yet exist in Genesis 3. We'll examine it further in both The Watchers and The Nephilim chapters.

Romans 5:12 and 14 reads, "[12] *Wherefore, as by one man sin entered into the world, and death by sin; and so death passed upon all men, for that all have sinned:* [14] *Nevertheless death reigned from Adam to Moses, even over them that had not sinned after the similitude of Adam's transgression, who is the figure of him that was to come.*"

According to scripture, it's not sin that is passed through the bloodline, but death. With this in mind, I propose a theory: Cain and Abel were born in the Garden of Eden. I won't say that I believe this theory, but I think it's plausible.

First, let's explore the issue of there being sinless bloodlines other than that of the Adams. Abel was killed, so he had no bloodline. Cain killed Abel, so he sinned in his own right.

There are clues in scripture that they may have been born in the garden. When Cain is born, Eve says that she had gotten a man from the Lord in Genesis 4:1, *"And Adam knew Eve his wife; and she conceived, and bare Cain, and said, I have gotten a man from the Lord."*

Contrast this to when Seth is born in Genesis 5:3, *"And Adam lived an hundred and thirty years, and begat a son in his own likeness, and after his image; and called his name Seth:"*

Adam begat Seth in his own likeness, after his image. This is interesting wording, almost as if Cain and Abel were not after Adam's image but after God's – just like Adam was in Genesis 1:27, *"So God created man in his own image, in the image of God created he him; male and female created he them."*

As for the timeline, Abel would have had to have been killed after the fall because he sacrificed animals. Animals were not killed in God's perfect Garden. Genesis 4 uses the phrase "in process of time it came to pass." This process of time would contain the fall in Genesis 4:3, *"And in process of time it came to pass, that Cain brought of the fruit of the ground an offering unto the Lord."*

There is one other hint that gives this theory credence. After the fall, God tells Eve that childbirth will be more painful. How would she have known unless she had already given birth? We read in Genesis 3:16, *"Unto the woman he said, I will greatly multiply thy sorrow and thy conception; in sorrow thou shalt bring forth children; and thy desire shall be to thy husband, and he shall rule over thee."*

You could say that in sequential verse order, Cain and Abel are born after the fall. It is true that the verses that say they were born come after the verses of the fall. But we have a verse before Adam and Eve are removed from the garden that belie the idea of the verses being sequential. Eve is called the mother of all living before they're removed from the garden in Genesis

3:20-21, *"And Adam called his wife's name Eve; because she was the mother of all living. Unto Adam also and to his wife did the Lord God make coats of skins, and clothed them."*

There is a parallel to Cain and Abel in Sumerian mythology. On clay tablets dating to 3000 B.C., is a Sumerian story titled **Cattle and Grain**. It talks about brother and sister deities who present their offerings to the gods:

> For Ashnan they establish a house,
> Plow and yoke they present to her.
> Lahar standing in his sheepfold,
> A shepherd increasing the bounty of the sheepfold is he;
> Ashnan standing among the crops,
> A maid kindly and bountiful is she.
>
> Abundance of heaven . . . ,
> Lahar and Ashnan caused to appear,
> In the assembly they brought abundance,

Long Lifespans

As seen in scripture, humans before the flood lived extremely long natural lives. There were many factors that allow for long lifespans. Some of those factors include the ice canopy, unmutated genes, eating unmutated diets, hyperbaric pressure and higher oxygen levels.

As mentioned in the chapter on the firmament, the theory of an ice canopy encapsulating Earth is a reasonable one. The ice canopy would have slowed the aging of mankind.

Certain electromagnetic frequencies cannot penetrate water. Visible light and radio waves travel through water, but ultraviolet and X-ray radiation do not. The lack of UV and X-rays would have enabled antediluvian man to live much longer lifespans.

Jonathan (left) is a tortoise still living today (2023) at 191 years old

The oldest known living land animal is a 191-year-old Seychelles giant tortoise named Jonathan. Jonathan lives on the island of St. Helena in the South Atlantic Ocean. His age is estimated, but a photograph was taken of him in 1886 at which time he was already at least 50 years old.

The reason that turtles live so long is likely a combination of factors. First, being cold-blooded they don't use a lot of energy to keep warm. Second, because they have a very slow metabolism, they don't need to eat as much, keeping the chemical reactions that cause aging to a minimum. Could their shells also keep their internal organs protected from UV light?

Long lifespans are likely due to the ice canopy as well as genetics. Evolution states that creatures are becoming more complex. Actually, our DNA is becoming more corrupt with every generation. We are in a state of devolution. While knowledge is increasing, our bodies are getting worse. When you make copies of a copy of a copy on a copying machine, the prints don't get better. They get worse. The same is true for our DNA.

Even in today's generations, we're seeing people with new ailments and allergies. Though gluten intolerance has been documented since A.D. 100, it's on the rise in the last 50 years. These ailments are also partially due to genetically modified foods causing things such as leaky gut. The green herbs that humans ate in the antediluvian world were full of nutrition because the generations were much closer to God's perfect creation. The soil was richer in minerals.

A recent BBC article, along with many others, report about our dwindling available farm land:

The dirt beneath our feet is getting poorer and on many farms worldwide, there is less and less of it. Without sufficient soil, our ability to grow food is threatened.

When you eat red meat, a small amount of the iron from that meat stays in your body. If you were to live much longer than 120 years, that iron in your body would kill you. God tells Adam that he can freely eat of the plants. God, in His wisdom, knew that if the Adams were to eat red meat (after the fall) that it would have prematurely killed them, because they lived for hundreds of years. After the flood, when men's lifespans dropped drastically, God tells Noah that man can now eat animals.

In Arthur Brown's book Methuselah's Secret, we find the disturbing and fascinating story of the Dickerson children. Reported on by the Windsor Star in the 1960s, the Dickersons were not allowed to have more than three children in their rental home. The parents hid their youngest three children in the attic for 15 years. The children were educated and fed but never allowed to go outside during the daytime.

Eventually, the children were discovered. They were amazingly small for their ages. Connie was 18, Gordon was 15 and Glenda was 13. Their mental capacity matched their age. Their lack of growth is likely due to the absence of UV light aging them.

The Dickerson children, Connie (18), Glenda (13), and Gordon (15)

In the Millennial kingdom, people will be considered children at 100 years old. We read in Isaiah 65:20, *"There shall be no more thence an infant of days, nor an old man that hath not filled his days: for the child shall die an hundred years old; but the sinner being an hundred years old shall be accursed."*

Because of the slower aging, antediluvian man was likely not sexually mature until around 100 years old. Notice that in scripture men were quite old (by our standards) before they had their first son. My timeline is taken from the Greek Septuagint:

Man	Age at Son's Birth
Adam	230 (3rd son)
Seth	205
Enosh	190
Cainan	170
Mahalalel	165
Jared	162
Enoch	165
Methuselah	187
Lamech	182
Noah	502

We see that Noah was 502 years old before he had his first child. Interestingly, early Jewish writers speculated that Noah knew of the coming judgment and thought it frivolous to bear children only to have them destroyed with the Earth. According to the book of Jasher 5:14-15, the Lord told Noah to marry Naamah and bear children, *"And the Lord said unto Noah, Take unto thee a wife, and beget children, for I have seen thee righteous before me in this generation. And thou shalt raise up seed, and thy children with thee, in the midst of the earth; and Noah went and took a wife, and he chose Naamah the daughter of Enoch, and she was five hundred and eighty years old."*

Deniers of the timeline of scripture will say that men couldn't have lived into their 900s, that scripture recorded their years as months so you must divide all ages by 12. If that's true, then according to the King James, Jared would have been 5.2 years old when he begat Enoch! That would be 13.5 years, according to the Septuagint.

Who did Cain marry? He likely married his sister. We see a lot of close-relation marriages in the antediluvian world and the years after the flood. Again, the human gene pool was perfect at creation. We have become so physically corrupt that genetic mutations are prominent in such relationships.

If you're from the United States, as I am, you likely frown on marriages between cousins. Today 10 percent of the world's marriages are between cousins. The Middle East has the highest rates of cousin marriage. In Saudi Arabia the number exceeds 70 percent.

In ancient Iranian mythology, the first man, Yima lived over 900 years in a golden age in which want, death, disease, aging and extreme temperatures were not present on the Earth. Interestingly he sounds a lot like Yama, the first man in Indian or Hindu mythology.

After the flood, we see a clear descension of the human lifespan.

The sun's UV rays had their effect on humans without the ice canopy overhead. According to the Septuagint timeline, these were the ages of the generations after the flood:

Man	Lifespan
Noah	950
Shem	600
Arphaxad	565
Shelah	460
Eber	504
Peleg	339
Reu	339
Serug	330
Nahor	208
Terah	205
Abraham	175

As with the declining ages, man's stature likely shrank too. Ron Wyatt claims to have found the graves of Noah and his wife. Apparently, Noah was 12 feet tall. This is plausible as most antediluvian life was roughly twice the size of their modern-day counterparts because of uncorrupted genetics and nutrition. Ron recovered a giant thumb bone at the Durupinar site. We will consider the Durupinar site as the likely location for Noah's ark in the chapter on the flood.

Advanced Ancient Technology

Were we once knuckle-dragging dimwits or were we created intelligent beings with the ability to speak and reason? Outside of the scriptures, there is fascinating evidence that mankind possessed technology superior to ours.

The Hindu Vedas are religious texts originating in ancient India around 1500 B.C. They contain the oldest scriptures of Hinduism. In the chapter on the firmament, we read a passage from the Rigveda. In the Mahabharata, we find an account of what can only be described as nuclear war.

In the Kurukshetra War, a projectile weapon, called Brahmastra, was used and produced explosions that leveled everything. Animals were engulfed in flames. Unborn babies died in the wombs of their mothers. Metal armor melted onto the skins of the warriors who wore it. Birds fell from the sky. The projectile was as bright as 10,000 suns. Written thousands of years ago, the Mahabharata reads:

> A single projectile charged with all the power in the Universe...An incandescent column of smoke and flame as bright as 10,000 suns, rose in all its splendor...it was an unknown weapon, an iron thunderbolt, a gigantic messenger of death which reduced to ashes an entire race... The corpses were so burned as to be unrecognizable. Their hair and nails fell out, pottery broke without any apparent cause, and the birds turned white... After a few hours, all foodstuffs were infected. To escape from this fire, the soldiers threw themselves into the river.

The Brahmastra would cause severe environmental damage for centuries. The land would become barren, rain would stop, vegetation would disappear and cracks would develop in the ground. People in the region would give birth to genetically defective children for generations.

Today in Rajasthan, India, radioactive ash covers a three-square-mile area. There is a very high rate of birth defects and cancer in the area. Scientists have unearthed a 10,000-year-old city that shows evidence of an atomic blast. The skeletons found are as radioactive as those at Hiroshima and Nagasaki.

At the nearby city of Mohenjo-Daro, the wall and foundations are fused together. An intense heat melted the walls, foundations and clay vessels which can only be explained by an atomic blast.

The Rig Veda documents vehicles capable of flying. They are described as flying palaces. One passage reads: *"jumping into space speedily with a craft using fire and water... containing twelve pillars, one wheel, three machines, 300 pivots, and 60 instruments."*

The Baghdad Battery is a clay pot discovered in 1936 in Iraq. It is on display at the National Museum in Iraq. It contains a copper cylinder that encases an iron rod. When filled with grape juice, the pot produces 1 volt of electricity. Scientists speculate that it was used to electroplate metals or for medicinal use.

The Antikythera Mechanism is a 1st-century B.C. mechanical device designed to calculate astronomical positions. It is the oldest known complex scientific calculator. It was recovered in 1901 from the Antikythera shipwreck.

Left: The Brahmastra | Right: Antikythera Mechanism Credit: Joyofmuseums, Wikimedia

Was all of this ancient technology wiped out in the great flood? What would happen to our vast knowledge in a global flood where only one family survived? How much would we lose simply by electricity being taken out in an EMP attack?

How did ancient man discover this knowledge? It was taught to them by the Watchers. We will study this in depth in the chapter on the Watchers.

CHAPTER 4

THE ANIMALS

What were the antediluvian animals like? Were they more intelligent before the fall or the flood? How many animal kinds have gone extinct since they stepped off the ark?

There exist animal species on our planet today that show intelligence. Ravens can multitask and solve difficult puzzles. Research has shown that they are smarter than great apes. Speaking of great apes, bonobos can learn to use sign language by watching videos. Parrots, of course, can mimic human speech.

We do have two instances in scripture of animals talking. The serpent in the Garden of Eden spoke to Eve. The serpent, I believe, was possessed by a fallen angel, but it still had the capability to speak. We read in Genesis 3:

1 Now the serpent was more subtil than any beast of the field which the Lord God had made. And he said unto the woman, Yea, hath God said, Ye shall not eat of every tree of the garden?

Also, Balaam's donkey speaks in Numbers 22:27-30,

"And when the ass saw the angel of the Lord, she fell down under Balaam: and Balaam's anger was kindled, and he smote the ass with a staff. And the Lord opened the mouth of the ass, and she said unto Balaam, What have I done unto thee, that thou hast smitten me these three times? And Balaam said unto the ass, Because thou hast mocked me: I would there were a sword in mine hand, for now would I kill thee. And the ass said unto Balaam, Am not I thine ass, upon which thou hast ridden ever since I was thine unto this day? was I ever wont to do so unto thee? and he said, Nay."

While the Lord opened the mouth of the donkey and enabled her to speak, in verse 30, the donkey is speaking of her life with Balaam. Of course, the Lord has the power to know the donkey's history, but it sure sounds like the donkey's intelligence dictating the words.

Were animals allowed to speak before the fall or the flood? It's possible since things between humans and animals drastically changed after God's covenant with creation after the flood. We read about it in Genesis 9:2-3, *"And the fear of you and the dread of you shall be upon every beast of the earth, and upon every fowl of the air, upon all that moveth upon the earth, and upon all the fishes of the sea; into your hand are they delivered. Every moving thing that liveth shall be meat for you; even as the green herb have I given you all things."*

Before the flood, the animals apparently didn't fear mankind. Wild animals today are skittish and only come near humans if they have grown up around them or have been fed by them. This innocence made the antediluvian animals easy prey for the corruption of the Nephilim which we will read about in a few chapters.

A common question is: Do animals go to Heaven? I don't know but I would guess that they don't. God never says in His word that the animals become a living soul as he does about the Adams. We read in Genesis 1:7, *"And the Lord God formed man of the dust of the ground, and breathed into his nostrils the breath of life; and man became a living soul."*

In Ecclesiastes 3:19-21, God says that the spirits of the animals go downward to the Earth while man's goes upward.

> "For that which befalleth the sons of men befalleth beasts; even one thing befalleth them: as the one dieth, so dieth the other; yea, they have all one breath; so that a man hath no preeminence above a beast: for all is vanity. All go unto one place; all are of the dust, and

all turn to dust again. Who knoweth the spirit of man that goeth upward, and the spirit of the beast that goeth downward to the earth?"

Apparently, beasts have spirits so that, at least, is a clue.

Large Animals

Many giant creatures existed in the past. Because of increased air pressure, genetically pure vegetation and long life spans, animals would grow to enormous sizes.

Insects breathe through spiracles in their skin, making their growth potential that much more than other creatures. Bigger insects are exactly what we find in the fossil record.

Instead of writing paragraphs for each of our giant, non-dinosaur scientific discoveries here's a list of what we've found:

16-inch-tall frog

Bird-eating spider with up to a 12-inch leg span

23-foot-tall rhinoceros – China 2015

Giant rodent, the size of a modern buffalo – Uruguay

9-foot-tall donkey

Buffalo skull with horns that span 12 feet

13-foot-tall turkey-like bird

7-foot-long beaver

18-inch cockroaches

2-foot-long grasshopper

8-foot centipedes – New Mexico 2005

3.5-foot tarantula

50-inch wingspan dragonfly

11-inch wingspan moth

15.5-inch-long snails

13-foot-long armadillos – South America

20-foot-tall sloths – South America

22-foot-long worms

45-foot-long snake called titanoboa – Columbia 2009

13-foot-tall camel

1-inch-long flea – China

12-foot-tall great apes – China

The increased air pressure in the antediluvian world would force more oxygen into the Earth's water. Because of the extra oxygen in the water, marine animals would grow to be enormous.

A one-inch shark's tooth indicates that its owner was about 14 feet long. The largest Megalodon tooth to have been found measured 7.48 inches. I've yet to see evolutionists explain the existence of over 100-foot long Megalodons in our fossil record.

Megalodon teeth

Fossils of extremely large water creatures have been found:

11-foot-wide oyster

50-foot-long squid

Octopi with 100-foot-long tentacles

25-foot-long manatees

8-foot-long sea scorpions – New York

12-foot claw spawn crabs

3-foot-long piranhas

16-foot-long turtles

3-foot-long shrimp - Morocco

The Chinese Giant Salamander is a "living fossil" and currently the largest amphibian on the planet. They can grow up to 6 feet in length. The longest living captive giant salamander lived for 52 years in the Amsterdam Zoo.

Dinosaurs & Dragons

Though they rolled a critical failure after the flood, dinosaurs thrived on the antediluvian Earth. Reptiles today can live extremely long lives. If dinosaurs lived longer than humans did before the flood, then some of them may have lived from creation all the way up to the flood. It's easy to see why these giant, long-lived killing machines dominated the planet.

Eighty-foot apatosauruses had nostrils the size of a modern-day horse. Dinosaurs have small lungs and small nostrils. An 80-foot dinosaur could not survive today. As we looked at in the chapter on the firmament, the antediluvian oxygen levels were 32 percent as opposed to 21 percent today.

Red blood cells carry oxygen through our bodies. Plasma is the liquid stream that the blood cells are carried in. With the increased atmospheric pressure in the antediluvian world, plasma could carry oxygen as well as the blood cells. These factors enabled dinosaurs with their tiny nostrils to thrive before the flood of Noah.

The recorded lifespan of humans was over 900 years before

the flood. Humans stopped growing at around 100 years in the antediluvian world. Reptiles never stop growing. With time, an uncorrupted diet, how big did the lizards get in the antediluvian world?

The word dinosaur means terrible lizard. You won't find the word dinosaur in the Bible because the word wasn't invented until 1841 by Sir Richard Owen. Before that time, they were known as dragons. In the Bible, they are called dragon, serpent, leviathan, cockatrice and behemoth.

Left: Hongshan jade dragon from Mongolia | Right: Dracorex skull. Credit: Kabacchi, Flickr

One of the earliest creatures to appear in the legends of ancient China is the dragon. It was worn on the robes of emperors, depicted on precious metals and referenced in literature and the performing arts dating back millennia. Despite being portrayed as a monster in most cultures, in China it's usually regarded as a benevolent creature. In one myth, Yu the Great (2070 B.C.) was helped by a dragon to manage floodwaters which were devastating his kingdom.

The earliest known depiction of a dragon is a c-shaped jade carving found in eastern Inner Mongolia. It belonged to the Hongshan culture between 4500 and 3000 B.C.

The dinosaur that looks most like what we would think of as a dragon is the dracorex. Though his body is more like that of a T. rex, his skull is something quite different.

We read about Behemoth in Job 40:15-18 and 23: *"Behold now behemoth, which I made with thee; he eateth grass as an ox. Lo now, his strength is in his loins, and his force is in the navel of his belly. He moveth his tail like a cedar: the sinews of his stones are wrapped together. His bones are as strong pieces of brass; his bones are like bars of iron....²³ Behold, he drinketh up a river, and hasteth not: he trusteth that he can draw up Jordan into his mouth."*

Some folks will say that these verses are describing a rhinoceros, hippopotamus or elephant but none of these creatures have a tail like a cedar. I believe that Job is describing a titanosaur, the heaviest land animal ever known.

Titanosaurs

Pterodactyls, though not dinosaurs themselves, lived alongside of dinosaurs and belonged to the pterosaur family. Fossils have been found with wingspans of 36 feet. We likely read about them in scripture. Isaiah 14:29 reads, *"Rejoice not thou, whole Palestina, because the rod of him that smote thee is broken: for out of the serpent's root shall come forth a cockatrice, and his fruit shall be a fiery flying serpent."*

Also in Isaiah, we read about a cockatrice den, and hatching cockatrice eggs. We read in Isaiah 59:5, *"They hatch cockatrice' eggs, and weave the spider's web: he that eateth of their eggs dieth, and that which is crushed breaketh out into a viper."*

Leviathan gets the honor of adorning this book cover. We read about him in Job 41:

> [1] Canst thou draw out leviathan with an hook? or his tongue with a cord which thou lettest down?... [14] Who can open the doors of his face? his teeth are terrible round about. [15] His scales are his pride, shut up together as with a close seal. [16] One is so near to another, that no air can come between them. [17] They are joined one to another, they stick together, that they cannot be sundered.

At this point leviathan sounds a lot like an alligator. We keep reading in Job 41:18-22, *"By his neesings a light doth shine, and his eyes are like the eyelids of the morning. Out of his mouth go burning lamps, and sparks of fire leap out. Out of his nostrils goeth smoke, as out of a seething pot or caldron. His breath kindleth coals, and a flame goeth out of his mouth. In his neck remaineth strength, and sorrow is turned into joy before him."*

We see that when he sneezes, light comes from his throat, his eyes shine and he breathes fire. Do I really believe that it was possible for dinosaurs to breathe fire? I do. Today, the bombardier beetle ejects a hot noxious chemical spray when threatened. It has two chemical compartments that when ejected, mix causing an explosion. Many dinosaur skulls have compartments that connect to their nasal passages. Some textbooks now say that some dinosaurs breathed fire.

We keep reading about leviathan in Job 41:

> [25] When he raiseth up himself, the mighty are afraid: by reason of breakings they purify themselves...[27] He esteemeth iron as straw, and brass as rotten wood. [28] The

arrow cannot make him flee: slingstones are turned with him into stubble...³⁰ Sharp stones are under him: he spreadeth sharp pointed things upon the mire. ³¹ He maketh the deep to boil like a pot: he maketh the sea like a pot of ointment. ³² He maketh a path to shine after him; one would think the deep to be hoary. ³³ Upon earth there is not his like, who is made without fear.

We also read about mighty leviathan in Psalm 74:12-14, *"For God is my King of old, working salvation in the midst of the earth. Thou didst divide the sea by thy strength: thou brakest the heads of the dragons in the waters. Thou brakest the heads of leviathan in pieces, and gavest him to be meat to the people inhabiting the wilderness."*

Notice that verse 14 mentions leviathan as having multiple heads. A hydra is a "mythological" multi-headed dragon. In Hesiod's ***Theogony***, written around 700 B.C., one of Hercules' 12 labors was to defeat the Lernean Hydra. Polycephaly is the condition of having more than one head. Can you guess which animal kind is most likely to being born with multiple heads? Reptiles and snakes are the most susceptible. The hydra may not be just a creature from mythology.

A turtle (turtles?) with polycephaly

Some folks, for whom I have great respect, will claim that leviathan is symbolic for Satan or another fallen angel. While possible, I believe that the Bible is literal when talking about leviathan. As with snakes or any creature, it's easy and logical to use them to visualize spiritual beings. Scripture does say that the sea is Leviathan's playground in Psalm 104:25-26, *"So is this great and wide sea, wherein are things creeping innumerable, both small and great beasts. There go the ships: there is that leviathan, whom thou hast made to play therein."*

Alas, Leviathan was to meet his doom. When the Lord is out to kill you, there is no fleeing. We read in Isaiah 27:1 that the Lord will kill Leviathan, *"In that day the Lord with his sore and great and strong sword shall punish leviathan the piercing serpent, even leviathan that crooked serpent; and he shall slay the dragon that is in the sea."*

In scripture, Rahab is the prostitute and heroine of the battle of Jericho. But the name Rahab is also used to describe a sea creature. We read in Isaiah 51:9, *"Awake, awake, put on strength, O arm of the Lord; awake, as in the ancient days, in the generations of old. Art thou not it that hath cut Rahab, and wounded the dragon?"*

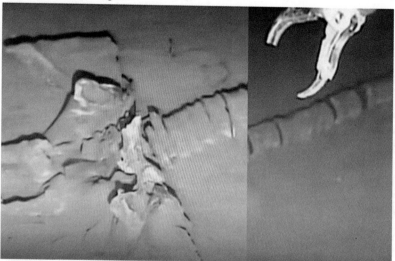

Unknown massive sea creature remains found in 2017

We also read of this Rahab in Psalm 89:10, *"Thou hast broken Rahab in pieces, as one that is slain; thou hast scattered thine enemies with thy strong arm."*

Though I can't be sure, Rahab in these passages may be referring to Leviathan.

According to *Journal News*, in 2017, a remote-operated vehicle pilot found the remains of a colossal, 100-foot, sea monster that matches no known creature.

The journal writes, "The diver who first spotted the mystery remains thinks that based on his previous discoveries, it could have far more ancient origins. 'The bone could be very ancient'..."

A video titled "The Remains of a Strange Creature Found On The Ocean Floor" can be viewed on YouTube. Could this giant sea creature be the remains of leviathan?

In the Masoretic Text, from which we get the King James Version of the Bible, Daniel is cast into the lion's den because he worshiped God. The apocrypha book of Daniel contains three more chapters in which we read about a living dragon. Not only did Daniel worship God, he refused to worship a real dragon. We read in Daniel 14:

> 23 There was a great dragon which the Babylonians revered. 24The king said to Daniel, "You cannot deny that this is a living god, so worship it." 25But Daniel answered, "I worship the Lord, my God, for he is the living God. 26Give me permission, O king, and I will kill this dragon without sword or club." "I give you permission," the king said. 27 Then Daniel took some pitch, fat, and hair; these he boiled together and made into cakes. He put them into the mouth of the dragon, and when the dragon ate them, he burst. "This," he said, "is what you revered." 28When the Babylonians heard this, they were angry and turned against the king. "The king has become a Jew," they said; "he has destroyed Bel, killed the dragon, and put the

priests to death." [29] They went to the king and demanded: "Hand Daniel over to us, or we will kill you and your family." [30] When he saw himself threatened with violence, the king was forced to hand Daniel over to them. [31] They threw Daniel into a lions' den, where he remained six days.

Daniel knew exactly how to kill this dragon. Daniel 1:3-4 says that Daniel was a knowledgeable lad understanding science, *"And the king spake unto Ashpenaz the master of his eunuchs, that he should bring certain of the children of Israel, and of the king's seed, and of the princes; Children in whom was no blemish, but well favoured, and skilful in all wisdom, and cunning in knowledge, and understanding science, and such as had ability in them to stand in the king's palace, and whom they might teach the learning and the tongue of the Chaldeans."*

A Chinese law book dating to 1611 B.C. lists the duties of a "Royal Dragon Feeder" who was responsible for throwing food into sacred ponds. Alexander the Great, came across a 100-foot-tall dragon while invading India in 330 B.C. The Roman author Claudius Aelianus in his work *On the Nature of Animals* tells of this encounter in book 15 chapter 21:

> When Alexander threw some parts of India into a commotion and took possession of others, he encountered among many other animals a serpent which lived in a cavern and was regarded as sacred by the Indians who paid it great and superstitious reverence. Accordingly Indians went to all lengths imploring Alexander to permit nobody to attack the serpent; and he assented to their wish. Now as the army passed by the cavern and caused a noise, the serpent was aware of it. (It has, you know, the sharpest hearing and the keenest sight of all animals.) And it hissed and snorted so violently that all were terrified and confounded. It was reported to measure 70 cubits although it was not visible in all its length, for it only put its head out. At any rate its eyes are said to have been the size of a large, round Macedonian shield.

Modern Dinosaurs

Did dinosaurs exist? They did. They do. Alligators and crocodiles are dinosaurs. Various evolutionist websites flail around about this difference or that difference that disqualifies alligators from being dinosaurs. It's an attempt to deny what is obvious so that they don't have to admit God exists and be shunned from mainstream science. In fact, evolutionists agree that alligators are older than dinosaurs. If that's true, why didn't the event that killed the dinosaurs kill the alligators?

I believe that alligators are some of the last surviving dinosaurs partly because they're aquatic. The biggest creatures on our planet today are sea creatures. The water helps block the sun's UV rays thus creating an environment where these species last longer and grow bigger.

There are a few reasons why dinosaurs have largely become extinct. The most likely reason is that they were eradicated because of the danger they posed to mankind. We have numerous stories of heroes setting off to slay dragons. One such story is that of Saint George.

Saint George was a soldier venerated in Christianity who slayed a dragon in Libya to rescue a princess. Though attributed to Saint George in the 11th century A.D., it likely was Saint Theodore Tiro who slew the dragon in the 9th century A.D. By that time, the dragons had grown so small that the one he slew was no bigger than a horse. Today, we have dozens of iconographies depicting saints George or Theodore slaying the dragon.

Beowulf is an Old English epic poem set in Scandinavia in the 6th century A.D. In the epic, Beowulf slays a dragon but is mortally wounded in the struggle. Interestingly, he, like King David, finds and uses a giant's sword at one point in the epic. Also of interest, Grendel, the monster of the epic is recorded as being a descendant of Cain, the cursed son of Adam.

Left: Saint George slays the dragon | Right: Pterosaur rock drawings. Credit: Tricia Simpson, Wikimedia

Tarascon is a modern-day city in France. It gets its name from the Tarasque dragon that was slain there in A.D. 48. Its description best matches that of an ankylosaurus. A statue to the Tarasque can be seen in the city today on the banks of the river.

The Buru is a 15-foot-long reptile described as looking like a cross between a crocodile and a snake. They lived in the Himalayan valley known as Rilo as recently as 1947. Their existence was confirmed by British zoologist Charles Stonor who got the same very detailed description from two different villages many miles apart. His studies are recorded in The Hunt for the Buru. Again, these long-lasting dinosaurs are partly aquatic.

In Black Dragon Canyon of Utah, a rock painting of a supposed pterosaur has been found. This is extremely problematic for militant evolutionists as fossils of pterosaurs were not found until 1784. There are many modern-day sightings of pterodactyls. Many of them have been reportedly seen flying the skies in Texas.

In 1856, workmen were digging a railway tunnel near St. Dizier France. When one large limestone boulder was split open, the workers were astonished to see a large winged creature tumble out. It fluttered its wings, let out a croaking noise and died. Their description matches that of a pterodactyl. Indeed, a paleontology student in the nearby town of Gray identified it as such. This story was reported in the Illustrated London News on February 9, 1856, on page 166.

The fact that the pterodactyl lived inside of a limestone boulder for thousands of years may seem impossible, but it's not the only story of a creature existing in statis. There are dozens of stories of frogs being enclosed in rocks and coming to life after the rocks are broken open. There are so many stories that the phenomenon is called "toad-in-the-hole."

Fish can be frozen for months and "come back to life" after the ice thaws. Alligators in colder climates stick their noses out of the water before they are frozen in place for winter. Once the ice thaws, they're back to normal. Can creatures, especially cold-blooded ones, live for long periods of time after being encased in rock?

A very compelling tale is that of the thunderbird of Tombstone. According to Native American lore, thunderbirds were huge birds of prey that created sounds of thunder when they flapped their wings together.

On April 26, 1890, the Tombstone Epitaph published an article about cowboys killing a thunderbird and taking a photo with it. Though hundreds report having seen the photograph in the 1970s and 1980s, no copies can be found dating to before the internet.

Between 1272 and 1295, the Venetian explorer Marco Polo visited Karazan just beyond Deadwind Pass. In his book, *The Travels of Marco Polo*, he describes dragons raised by a Chinese emperor to pull his chariots:

> Here are seen huge serpents, ten paces in length, and ten spans in the girt of the body. At the fore-part, near the

head, they have two short legs, having three claws like those of a tiger, with eyes larger than a fourpenny loaf (pane da quattro denari) and very glaring. The jaws are wide enough to swallow a man, the teeth are large and sharp, and their whole appearance is so formidable, that neither man, nor any kind of animal, can approach them without terror.

In 1571, Spanish conquistadors discovered stones (the Ica Stones) in Peru—over 500 of which feature carvings of humans with dinosaurs even some humans riding dinosaurs. One evolutionist on Wikipedia laughably says that this image "allegedly" depicts dinosaurs.

You'll notice skin texture on the depicted dinosaurs. It took archaeologists until 1997 to find fossilized dinosaur skin in Bolivia. How would the ancients know what the skin looked like unless they had seen them?

Ica Stones depicting dinosaurs. Credit: Brattarb, Wikipedia

In the 1940s, over 33,700 ceramic figurines were discovered in Mexico. They depict hundreds of different dinosaurs and extinct Ice Age animals. Some of the figures have humans riding on the dinosaurs. Scientists agree that these figurines are over 3,000 years old.

The Acambaro Figurines. Credit: Brattarb, Wikimedia

The Loch Ness monster has been seen for hundreds of years in Scotland. The monster was sighted 52 times alone in 1933 when a road was built alongside the water. In Canada, a similar creature has been spotted and is called Ogopogo.

In 1977, Japanese fishermen netted the corpse of a strange creature off the coast of New Zealand. It was 32-feet-long and weighed 4,000 pounds. There are no available images for licensing so I can't include a picture here. Search for "zuiyo-maru" and you'll see images.

In the mid 2000s, scientists were astonished to find red blood cells and soft tissue on a fossilized leg bone of a Tyrannosaurus rex. After all, soft tissues are the first materials to disappear during the fossilization process, especially after millions of years of fossilization. Sergio Bertazzo is a materials scientist at Imperial College London. When he saw soft tissue on a fossilized theropod claw, he was shocked. He said, "One morning, I turned on the microscope, increased the magnification, and thought, 'Wait – that looks like blood!'"

After the discovery, their reaction wasn't to say that the theropod may only have died a few thousand years ago. Their reaction was to say that soft tissue can survive for 75 million years. Anything to deny God's existence.

Did dinosaurs and man coexist? Man has a natural desire to be free from authority. There will never be enough evidence for a militant evolutionist to admit that God is real. These stories confirm what God says in His Word – man and dinosaurs were created in the same week and therefore coexisted.

CHAPTER 5

THE WATCHERS

Elohim created a fantastical planet. Why then did He send a great flood to destroy it? All flesh including mankind and the animals had become corrupt according to Genesis 6:

> [5]And God saw that the wickedness of man was great in the earth, and that every imagination of the thoughts of his heart was only evil continually. [6]And it repented the Lord that he had made man on the earth, and it grieved him at his heart. [7]And the Lord said, I will destroy man whom I have created from the face of the earth; both man, and beast, and the creeping thing, and the fowls of the air; for it repenteth me that I have made them....[11]The earth also was corrupt before God, and the earth was filled with violence. [12]And God looked upon the earth, and, behold, it was corrupt; for all flesh had corrupted his way upon the earth.

Calculated by the genealogy from Adam to Noah, according to the Greek Septuagint, there was a period of about 2,256 years between Adam's creation or fall and the flood. In that timeframe, what caused man to be so wicked? Why did God repent that he had made the animals? It was the Watchers (fallen angels) and their offspring the Nephilim that had corrupted the Earth. We read about their entrance four verses earlier in Genesis 6:

> [1]And it came to pass, when men began to multiply on the face of the earth, and daughters were born unto them, [2]That the sons of God saw the daughters of men that they were fair; and they took them wives of all which they chose. [3]And the Lord said, My spirit shall not always strive with man, for that he also is flesh: yet his days shall be an hundred and twenty years. [4]There were giants in

the earth in those days; and also after that, when the sons of God came in unto the daughters of men, and they bare children to them, the same became mighty men which were of old, men of renown.

These "sons of God" are not the human line of Seth as some folks will claim. There are many instances in scripture telling us so. The Hebrew for "sons of God" is *bə·nê hā·'ĕ·lō·hîm*. We also read about the bene ha Elohim in Job 2:1-2, *"Again there was a day when the sons of God came to present themselves before the Lord, and Satan came also among them to present himself before the Lord. And the Lord said unto Satan, From whence comest thou? And Satan answered the Lord, and said, From going to and fro in the earth, and from walking up and down in it."*

As we can see, these sons of God are <u>in</u> Heaven in Job 2 and therefore cannot be humans from the line of Seth. Schofield uses Matthew 22:30 to say that these cannot be angels, *"For in the resurrection they neither marry, nor are given in marriage, but are as the angels of God in heaven."*

These angels in Genesis 6 are neither God's angels nor in Heaven. They rebelled against Elohim and chose a mortal life. We read about this in Jude 6:1, *"And the angels which kept not their first estate, but left their own habitation, he hath reserved in everlasting chains under darkness unto the judgment of the great day."*

The idea of angels taking on human form and touching people should not be foreign to us as Christians. There are many times in scripture where angels touch and even eat with humans. Two angels physically deliver Lot and his family out of Sodom in Genesis 19:

> ¹⁰ But the men put forth their hand, and pulled Lot into the house to them, and shut to the door....¹⁵ And when the morning arose, then the angels hastened Lot, saying,

Arise, take thy wife, and thy two daughters, which are here; lest thou be consumed in the iniquity of the city. [16] And while he lingered, the men laid hold upon his hand, and upon the hand of his wife, and upon the hand of his two daughters; the Lord being merciful unto him: and they brought him forth, and set him without the city.

In Acts 12:7, the angel of the Lord smote Peter while he was snoozing, "And, behold, the angel of the Lord came upon him, and a light shined in the prison: and he smote Peter on the side, and raised him up, saying, Arise up quickly. And his chains fell off from his hands."

In 1 Kings 19:5, an angel touched Elijah to wake him up, *"And as he lay and slept under a juniper tree, behold, then an angel touched him, and said unto him, Arise and eat."*

In Genesis 181-2, 8, God and two angels eat with Abraham, *"And the Lord appeared unto him in the plains of Mamre: and he sat in the tent door in the heat of the day; And he lift up his eyes and looked, and, lo, three men stood by him: and when he saw them, he ran to meet them from the tent door, and bowed himself toward the ground,... And he took butter, and milk, and the calf which he had dressed, and set it before them; and he stood by them under the tree, and they did eat."*

In Genesis 32:24, 29-30, Jacob wrestles with God, *"And Jacob was left alone; and there wrestled a man with him until the breaking of the day.... And Jacob asked him, and said, Tell me, I pray thee, thy name. And he said, Wherefore is it that thou dost ask after my name? And he blessed him there. And Jacob called the name of the place Peniel: for I have seen God face to face, and my life is preserved."*

Considering these verses, it should not be hard to believe scripture when it says that angels entered our physical realm and interacted with humans. It also should not be hard to understand that fallen angels slept with human women.

In scripture, there is a lot of information left out about what really took place in Genesis 6. We find the whole story in the apocryphal book of Enoch.

Again, I do not believe that the book of Enoch was inspired by God. I believe the 66 books of the Old and New Testaments are inspired by God. I do believe that the book of Enoch is correct history. I arrived at this conclusion after years of study and fitting puzzle pieces together. Whatever you believe, it's imperative that you filter anything that I write from an extrabiblical source through the word of God.

The Watchers from their celestial perches observed human women and lusted after them. Two-hundred Watchers led by Semjâzâ decided to rebel against Elohim and take on human form and live among mankind. During the days of Jared, the fifth from Adam, the Watchers descended onto Mount Hermon. Jared was born in 4594 B.C .and died in 3632 B.C. Enoch, Jared's son, knew of the Watchers and was taken up to Heaven in 4067 B.C. That puts the Watchers arriving between 4594 B.C. and 4067 B.C. This is 960 years after Adam's fall if you agree with the Greek Septuagint timeline as I do.

We read about the fall of the Watchers in Enoch 6:

> [1.] And it came to pass when the children of men had multiplied that in those days were born unto them beautiful and comely daughters. [2.] And the angels, the children of the heaven, saw and lusted after them, and said to one another: 'Come, let us choose us wives from among the children of men and beget us children.' [3.] And Semjâzâ, who was their leader, said unto them: 'I fear ye will not indeed agree to do this deed, and I alone shall have to pay the penalty of a great sin.' [4.] And they all answered him and said: 'Let us all swear an oath, and all bind ourselves by mutual imprecations not to abandon this plan but to do this thing.' [5.] Then sware they all together and bound themselves by mutual imprecations upon it.

6. And they were in all two hundred; who descended in the days of Jared on the summit of Mount Hermon, and they called it Mount Hermon, because they had sworn and bound themselves by mutual imprecations upon it. 7. And these are the names of their leaders: Sêmîazâz, their leader, Arâkîba, Râmêêl, Kôkabîêl, Tâmîêl, Râmîêl, Dânêl, Êzêqêêl, Barâqîjâl, Asâêl, Armârôs, Batârêl, Anânêl, Zaqîêl, Samsâpêêl, Satarêl, Tûrêl, Jômjâêl, Sariêl. 8. These are their chiefs of tens.

Verse 6 says that 200 Watchers descended. Verses 7 and 8 say that 190 descended. Penemue is one more chief mentioned in chapter 69 bringing the tally up to 200. Interestingly, in the Sumerian creation story, which we talked about in the chapter on the creation, says that the Anunnaki came to Earth in groups of 200.

We have evidence in 1 Corinthians 15:39-40 that the bodies of human women had their own splendor that angels' bodies don't and vice versa, *"All flesh is not the same flesh: but there is one kind of flesh of men, another flesh of beasts, another of fishes, and another of birds. There are also celestial bodies, and bodies terrestrial: but the glory of the celestial is one, and the glory of the terrestrial is another."*

The story of the Watchers lusting after human women affected women in the New Testament and indeed in some churches today. We read in 1 Corinthians 11:5-6, *"But every woman that prayeth or prophesieth with her head uncovered dishonoureth her head: for that is even all one as if she were shaven. For if the woman be not covered, let her also be shorn: but if it be a shame for a woman to be shorn or shaven, let her be covered."*

Paul continues in verses 10-11, *"For this cause ought the woman to have power on her head because of the angels. Nevertheless neither is the man without the woman, neither the woman without the man, in the Lord."*

1 Corinthians clearly tells us that the angels are the cause for women having to wear head coverings. It was so that the angels would not see their beauty and lust after them.

We see that the Watchers knew that God would punish them with a mortal death for their rebellion. It is speculated that one reason the Watchers wanted to enter our physical timeline was to escape the judgment even if only for a 500-year reprieve. Since time is a physical construct, and God always has been and always will be, I believe that the spiritual realm exists outside of our timeline. If that's the case, the cross and the Great Day of Judgment has already happened, will happen, and is happening in the spiritual realm.

I know that the more powerful spiritual beings, if not all of them, can appear anywhere on the Earth at any given moment they choose. This was proven to me by a story told by Pastor Greg Patten. You should check out his work at gregpatten.com. Pastor Patten has dealt with demonic possession for over 40 years. To me, he is a modern-day Job. Satan has attempted to destroy Greg's life, but Greg has never once cursed God. Pray for Greg Patten.

Pastor Patten has many chilling stories of his encounters while trying to help people. In one such story, a demon was getting agitated with Greg's rebuke, so the demon told him to hold on while he goes to get Moloch. Within seconds, Greg's house was shaking to its foundation. That story tells me that at least powerful demons can appear wherever they want in an instant.

Forbidden Knowledge and Corruption

The Watchers corrupted humankind by teaching them forbidden knowledge and intermingling with human women. They corrupted man's innocence and bloodline. We read about some of the forbidden knowledge taught to mankind in Enoch chatper 7:1, *"And all the others together with them took unto*

*themselves wives, and each chose for himself one, and they
began to go in unto them and to defile themselves with them,
and they taught them charms and enchantments, and the cutting
of roots, and made them acquainted with plants."*

We continue the list in Enoch 8:

1. And Aʒâʒêl taught men to make swords, and knives, and
shields, and breastplates, and made known to them the
metals (of the earth) and the art of working them, and
bracelets, and ornaments, and the use of antimony, and
the beautifying of the eyelids, and all kinds of costly
stones, and all colouring tinctures. 2. And there arose
much godlessness, and they committed fornication, and
they were led astray, and became corrupt in all their
ways. Semjâʒâ taught enchantments, and root-cuttings,
Armârôs the resolving of enchantments, Barâqîjâl,
(taught) astrology, Kôkabêl the constellations, Eʒêqêêl
the knowledge of the clouds, (Araqiêl the signs of the
earth, Shamsiêl the signs of the sun), and Sariêl the
course of the moon. And as men perished, they cried, and
their cry went up to heaven . . .

We continue the list in Enoch 69:

8. And the name of the fourth is Penemue; this one showed
the sons of men the bitter and the sweet and showed them
all the secrets of their wisdom. 9. He taught men the art of
writing with ink and paper, and through this many have
gone astray, from eternity to eternity, and to this day....
12. And the name of the fifth is Kasdeyae; this one showed
the sons of men all the evil blows of the spirits and of
the demons, and the blows that attack the embryo in the
womb so that it miscarries. And the blows that attack the
soul: the bite of the serpent. And the blows that occur at
midday, and the son of the serpent – who is strong.

From these verses, we compile a list of forbidden knowledge taught to mankind by the Watchers: charms, enchantments, herbalism, bladed weapons, shields, armor, metallurgy, jewelry, elements, makeup, precious stones, painting, astrology, constellations, clouds, signs of the Earth, signs of the sun, signs of the moon, writing, demonology, abortion, snake bites and spiritualism. In Enoch 9:6, we see how important this knowledge was, *"Thou seest what Aẑâẑêl hath done, who hath taught all unrighteousness on earth and revealed the eternal secrets which were (preserved) in heaven, which men were striving to learn."*

It's not just the Watchers teaching knowledge. In Judges 3:1-2, interestingly, God wants to teach His own people how to wage war, *"Now these are the nations which the Lord left, to prove Israel by them, even as many of Israel as had not known all the wars of Canaan; Only that the generations of the children of Israel might know, to teach them war, at the least such as before knew nothing thereof."*

There is one passage in scripture that I have found that mentions knowledge taught to mankind by the angels. We read in Revelation 21:17 about a man measuring the walls of Heaven in John's vision, *"And he measured the wall thereof, an hundred and forty and four cubits, according to the measure of a man, that is, of the angel."*

Here we read that man learned the cubit measurement from the angels. The cubit was the unit of measurement for most ancient cultures. It's the measurement from the tip of your middle finger to your elbow. Here a man is using it to measure the walls of Heaven.

One thing that archaeologists agree on is that an advanced civilization, the Sumerians, appeared suddenly on the planet without any technological lead up about 50,000 years ago (their timeline) with advanced science, mathematics, astronomy, a legal system, agriculture and writing. How could this happen?

The Sumerians themselves tell us how they got this knowledge. They say that Oannes, the half fish-half-man walked out of the ocean and told them that they were going to worship him and he would teach them advanced knowledge including writing. We talk extensively about Oannes and his teachings in my previous book, *Ancient Cities and the gods Who Built Them*.

It was also at this same time that Neanderthals appear in the archeological record, according to secular scientists. In 1985, archaeologists excavating the Cueva de los Aviones in Murcia found seashell necklaces and oyster shells containing orange, red and yellow makeup. These things belonged to the Neanderthals that lived there. We'll talk a bit more about Neanderthals in the chapter on the Nephilim.

The Watchers corrupted mankind with forbidden knowledge. If Adam and Eve had never eaten of the Tree of the Knowledge of Good and Evil, would the Watchers have taught humans that knowledge, as well?

In Sumerian mythology, the Anunnaki are depicted as astronauts that taught heavenly knowledge to mankind. The name Anunnaki is derived from An, the Sumerian god of the sky. The name Anunnaki means princely offspring.

Rock Drawings in Valcamonica, Italy look suspiciously like they come in peace. Credit: Luca Giarelli, Wikipedia

Mankind wasn't the only thing that the Watchers corrupted. We read about the Watchers corruption of the Earth through their offspring, the Nephilim, in Enoch 7:

> 2. And they became pregnant, and they bare great giants, whose height was three thousand ells: 3. Who consumed all the acquisitions of men. And when men could no longer sustain them, 4. the giants turned against them and devoured mankind. 5. And they began to sin against birds, and beasts, and reptiles, and fish, and to devour one another's flesh, and drink the blood. 6. Then the earth laid accusation against the lawless ones.

Notice that the giants began to devour mankind and drink the blood. Scripture mentions the giants eating mankind in Numbers 13:32-33, *"And they brought up an evil report of the land which they had searched unto the children of Israel, saying, The land, through which we have gone to search it, is a land that eateth up the inhabitants thereof; and all the people that we saw in it are men of a great stature. And there we saw the giants, the sons of Anak, which come of the giants: and we were in our own sight as grasshoppers, and so we were in their sight."*

Why were the giants drinking the blood? Why does anyone drink blood? God tells us in scripture that life is in the blood in Leviticus 17:

> 10 And whatsoever man there be of the house of Israel, or of the strangers that sojourn among you, that eateth any manner of blood; I will even set my face against that soul that eateth blood, and will cut him off from among his people. 11For the life of the flesh is in the blood: and I have given it to you upon the altar to make an atonement for your souls: for it is the blood that maketh an atonement for the soul. 12 Therefore I said unto the children of Israel, No soul of you shall eat blood, neither shall any stranger that sojourneth among you eat blood.

...[14] For it is the life of all flesh; the blood of it is for the life thereof: therefore I said unto the children of Israel, Ye shall eat the blood of no manner of flesh: for the life of all flesh is the blood thereof: whosoever eateth it shall be cut off.

Blood consists mostly of protein and water. It is rich with iron and vitamin D. While it may be logical to consume such a nutritious meal for someone who doesn't live with modern-day amenities, God strictly forbids it.

How does this relate to some Christian denominations belief that the wine of the sacrament of the Eucharist (communion) turns into the real blood of Jesus Christ? Scripture even goes so far as to say that the blood of specific animals cannot take away sins in Hebrews 10:4, *"For it is not possible that the blood of bulls and of goats should take away sins."*

Spiritual beings seem to be capable of experiencing a different interaction with blood than do humans, according to Genesis 4:9-11, *"And the Lord said unto Cain, Where is Abel thy brother? And he said, I know not: Am I my brother's keeper? And he said, What hast thou done? the voice of thy brother's blood crieth unto me from the ground. And now art thou cursed from the earth, which hath opened her mouth to receive thy brother's blood from thy hand."*

We find an interesting verse in Hebrews 12:24 that is relevant, *"And to Jesus the mediator of the new covenant, and to the blood of sprinkling, that speaketh better things than that of Abel."*

It seems that the shedding of Abel's blood meant something unseen by man. Abel's was the first blood of mankind to be shed. I won't take the space to list the dozens of verses in the New Testament that say that salvation is through the blood of Christ.

From these passages, we see that there is more to blood than just the physical aspects. In a spiritual context, life is in the blood. Could there be some power that a supernatural being could obtain through the consumption of another's life source?

The Mixing of DNA

A chimera is an individual whose body is composed of cells that are genetically distinct, as if they are from different individuals or creatures. In mythologies, we see many strange beings. The Egyptian, Greek, Norse and Roman gods are super-humans with super-powers often depicted as human and animal hybrids.

In Mesopotamia, Pazuzu, the king of demons, has the body of a man, the head of a lion, the talons of an eagle, wings and a scorpion's tail. Ganesha, is a Hindu god with the body of a man and the head of an elephant. In Greek mythology, Pan is the god of nature with the torso of a man and the horns and legs of a goat.

Mermaids are in most mythologies. Jengu is a half fish-half -woman in African cultures. Sirena are mermaids from Philippine folklore. Dagon is a merman worshiped in Mesopotamian and Assyrian mythology.

Harpies are Greco-Roman mythological bird women. Cecaelia is a half-female-half-octopus. Hundreds of cultures, religions, mythologies and legends tell of these half human-half animal deities.

Human/animal hybrids are found in ancient cave paintings, as well. Cave paintings with chimeras are found in Australia's Arnhem Land, France's Lascaux cave and India's Bhimbetka Rock Shelters.

While it's hard to decipher the true visage of a creature from just the bones, cave paintings give us insight on what they may have looked like. Cave paintings in Arnhem Land depict a long ago extinct marsupial lion with stripes called a Thylacoleo.

Not only do cave paintings reveal extinct and strange creatures, but they can also show that some animals lived where we don't see them today. In the Amazon, is an 8-mile-long stone canvas depicting thousands of paintings of animals from thousands of years ago, including elephants.

Are these hundreds of named beings from mythology all fictional characters created with a Tolkienesque mindset, or might there be some truth involved? Scripture says that one reason for the flood is that the whole Earth had become corrupt. We read in Genesis 6:7, 12, *"And the Lord said, I will destroy man whom I have created from the face of the earth; both man, and beast, and the creeping thing, and the fowls of the air; for it repenteth me that I have made them. ...12 And God looked upon the earth, and, behold, it was corrupt; for all flesh had corrupted his way upon the earth."*

The book of Enoch records some of the corruption by the Nephilim in chapter 7:4-5, "the giants turned against them and devoured mankind. And they began to sin against birds, and beasts, and reptiles, and fish, and to devour one another's flesh, and drink the blood."

Were the Watchers and/or the Nephilim mixing the DNA of Elohim's creation? I can't say exactly what Enoch means when he says that the Nephilim were sinning against the animals, but it doesn't sound good. I speculate that they were mixing animal and human DNA – thus the ancient stories of human/animal hybrid demigods.

When Jesus says that as it was in the days of Noah, so shall it be at the coming of the Son of Man, does he mean that in addition to great wickedness, we will see some of these chimeras, too? We read an interesting passage about the Earth's future in Revelation 13:14-15, *"And deceiveth them that dwell on the earth by the means of those miracles which he had power to do in the sight of the beast; saying to them that dwell on the earth, that* they should make an image to the beast, which had the wound by a sword, and did live. And he had power to give life unto the image of the beast, that the image of the beast should both speak, and cause that as many as would not worship the image of the beast should be killed."

What is this image of the beast? Why is the Antichrist able to breathe life into it? Is it a clone or chimera? Does God give us a hint about the future being like the days of Noah?

In our current age of depravity and technology, we see reports of the mixing of DNA happening. In an August 21, 2019, Fox News article, Tucker Carlson reports on China's creation of human/monkey hybrids in the attempt to grow organs for human transplants. Genesis 11:6 sums the situation up nicely, *"And the Lord said, Behold, the people is one, and they have all one language; and this they begin to do: and now nothing will be restrained from them, which they have imagined to do."*

In April of 2021, the U.S. Senate passed the Endless Frontier Act to help the U.S. better compete with Chinese innovation. It would have allowed the bioengineering and experimentation of human-animal hybrid species.

In a report about the legislation, Lara Logan said, "Chimeric research is a Pandora's box from Hell that should not be opened. The scientific community spurred on by their own hubris may very well open it using taxpayer dollars."

While this effort didn't gain enough support to pass the U.S. House of Representatives, it did garner enough votes to pass the U.S. Senate. We are very close to taking another step in corrupting Elohim's postdiluvian Earth.

The Judgment of The Watchers

The Watchers utterly corrupted Elohim's creation and the bloodline of mankind except for Noah and his sons. The Watchers and their offspring had sinned against the animals.

The cries of men went up to Heaven.

It's at this point that four of the archangels see what's happening to Elohim's creation. We read in Enoch 9:

1. And then Michael, Uriel, Raphael, and Gabriel looked down from heaven and saw much blood being shed upon the earth, and all lawlessness being wrought upon the earth. 2. And they said one to another: 'The earth made without inhabitant cries the voice of their crying up to the gates of heaven. 3And now to you, the holy ones of heaven, the souls of men make their suit, saying, "Bring our cause before the Most High.".' 4. And they said to the Lord of the ages: 'Lord of lords, God of gods, King of kings, (and God of the ages), the throne of Thy glory (standeth) unto all the generations of the ages, and Thy name holy and glorious and blessed unto all the ages! 5. Thou hast made all things, and power over all things hast Thou: and all things are naked and open in Thy sight, and Thou seest all things, and nothing can hide itself from Thee. 6. Thou seest what Azâzêl hath done, who hath taught all unrighteousness on earth and revealed the eternal secrets which were (preserved) in heaven, which men were striving to learn: 7. And Semjâzâ, to whom Thou hast given authority to bear rule over his associates. 8. And they have gone to the daughters of men upon the earth, and have slept with the women, and have defiled themselves, and revealed to them all kinds of sins. 9. And the women have borne giants, and the whole earth has thereby been filled with blood and unrighteousness. 10. And now, behold, the souls of those who have died are crying and making their suit to the gates of heaven, and their lamentations have ascended: and cannot cease because of the lawless deeds which are wrought on the earth. 11. And Thou knowest all things before they come to pass, and Thou seest these things and Thou dost suffer

them, and Thou dost not say to us what we are to do to them in regard to these.'

Let's take a synonymous rabbit trail about the angels before we bring it back to the fate of the rebellious Watchers. Seven archangels are mentioned in ancient texts including Enoch 20. The archangels mentioned are Michael, Raphael, Gabriel, Uriel, Sariel, Raguel, and Remiel. We read in Enoch 20:

> 1. And these are the names of the holy angels who watch.
> 2. Uriel, one of the holy angels, who is over the world and over Tartarus. 3. Raphael, one of the holy angels, who is over the spirits of men. 4. Raguel, one of the holy angels who †takes vengeance on† the world of the luminaries. 5. Michael, one of the holy angels, to wit, he that is set over the best part of mankind and over chaos. 6. Saraqâêl, one of the holy angels, who is set over the spirits, who sin in the spirit. 7. Gabriel, one of the holy angels, who is over Paradise and the serpents and the Cherubim. 8. Remiel, one of the holy angels, whom God set over those who rise.

Only Michael and Gabriel are mentioned in scripture. Raphael is mentioned in the apocryphal Old Testament Book of Tobit.

Revelation 12:7, *"And there was war in heaven: Michael and his angels fought against the dragon; and the dragon fought and his angels"*

Jude 1:9, *"Yet Michael the archangel, when contending with the devil he disputed about the body of Moses, durst not bring against him a railing accusation, but said, The Lord rebuke thee."*

Luke 1:19, 26, *"And the angel answering said unto him, I am Gabriel, that stand in the presence of God; and am sent to speak unto thee, and to shew thee these glad tidings....26 And in the sixth month the angel Gabriel was sent from God unto a city of Galilee, named Nazareth"*

Daniel 8:16, *"And I heard a man's voice between the banks of Ulai, which called, and said, Gabriel, make this man to understand the vision."*

Daniel 9:21, *"Yea, whiles I was speaking in prayer, even the man Gabriel, whom I had seen in the vision at the beginning, being caused to fly swiftly, touched me about the time of the evening oblation."*

Tobit 12:15-17, *"I am Raphael, one of the seven angels who stand and serve before the Glory of the Lord." Greatly shaken, the two of them fell prostrate in fear. But Raphael said to them: "Do not fear; peace be with you! Bless God now and forever."*

Back to the fate of the rebellious Watchers ... God gives four of His archangels each a different task. To Uriel, God gives the task of informing Noah of the impending doom. We will read about this in the book of Enoch in a couple of pages.

God ascribes the different types of corruption to specific Watchers and assigned different archangels to capture them. To Azâzêl God ascribes the corruption of mankind through forbidden knowledge and sends Raphael to bind him. We read in Enoch 10:

> [4.] And again the Lord said to Raphael: 'Bind Aʒâʒêl hand and foot, and cast him into the darkness: and make an opening in the desert, which is in Dûdâêl, and cast him therein. [5.] And place upon him rough and jagged rocks, and cover him with darkness, and let him abide there for ever, and cover his face that he may not see light. [6.] And on the day of the great judgement he shall be cast into the fire. And heal the earth which the angels have corrupted, and proclaim the healing of the earth, that they may heal the plague, and that all the children of men may not perish through all the secret things that the Watchers have disclosed and have taught their sons. [8.] And the whole earth has been corrupted through the works that were taught by Aʒâʒêl: to him ascribe all sin.

To the Nephilim, the offspring of the angels, God ascribes destroying humans and sends Gabriel to destroy them. We read in Enoch 10:9-10, "And to Gabriel said the Lord: 'Proceed against the bastards and the reprobates, and against the children of fornication: and destroy the children of fornication and the children of the Watchers from amongst men and cause them to go forth: send them one against the other that they may destroy each other in battle: for length of days shall they not have. And no request that they (i.e. their fathers) make of thee shall be granted unto their fathers on their behalf; for they hope to live an eternal life, and that each one of them will live five hundred years.' "

While the flood did finish off the antediluvian Nephilim, there likely weren't many giants left to be destroyed. Before the flood happens, Gabriel causes them to fight against and kill each other.

To Semjâzâ God ascribes defiling human women and sends Michael to bind him. We read in Enoch 10:11-12, "And the Lord said unto Michael: 'Go, bind Semjâzâ and his associates who have united themselves with women so as to have defiled themselves with them in all their uncleanness. And when their sons have slain one another, and they have seen the destruction of their beloved ones, bind them fast for seventy generations in the valleys of the earth, till the day of their judgement and of their consummation, till the judgement that is for ever and ever is consummated."

Why do I believe that the book of Enoch gets the events right? Because We have many places in scripture that mention the rebellious angels being cast into the prison. We read Jude 6:1 earlier in the chapter, but it's worth repeating here, *"And the angels which kept not their first estate, but left their own habitation, he hath reserved in everlasting chains under darkness unto the judgment of the great day."*

We also read about these spirits in prison in 1 Peter 3:18-20, "*For Christ also hath once suffered for sins, the just for the unjust, that he might bring us to God, being put to death in the flesh, but quickened by the Spirit: By which also he went and preached unto the spirits in prison; Which sometime were disobedient, when once the longsuffering of God waited in the days of Noah, while the ark was a preparing, wherein few, that is, eight souls were saved by water.*"

After Jesus died on the cross, while His physical body was dead, His spirit goes down into the abyss and He "preached" to the spirits that made trouble in the days of Noah. The Greek word for preached is ***kērussō*** which means to preach or to proclaim. It's more likely that Jesus proclaimed His victory instead of preached to the Watchers in prison.

Again, as I stated earlier, the Bible is my foundation. In trying to solve a mystery, we put puzzle pieces together. When the Bible brings together pieces, I count those as truth. There is a hole in the puzzle that this angelic story from the book of Enoch fits perfectly. As long as it doesn't contradict scripture, I will keep an open mind as I study the mysteries of our universe.

Let's keep looking at scripture and linking to the story in the book of Enoch. We know the book of Isaiah as the biggest source of prophecy of the coming Messiah. As we just read in 1 Peter 3, Isaiah 24:21-22 foretells of the Jesus' visit with these fallen angels, "*And it shall come to pass in that day, that the Lord shall punish the host of the high ones that are on high, and the kings of the earth upon the earth. And they shall be gathered together, as prisoners are gathered in the pit, and shall be shut up in the prison, and after many days shall they be visited.*"

2 Peter 2:4-5 reads, "*For if God spared not the angels that sinned, but cast them down to hell, and delivered them into chains of darkness, to be reserved unto judgment; And spared not the old world, but saved Noah the eighth person, a preacher*

of righteousness, bringing in the flood upon the world of the ungodly;"

While the book of Enoch is not seen as a reliable source by many Christians, as we've seen, there is a lot of scripture that confirms some of the events that take place in it.

We will take an in-depth look at the great flood in a later chapter so here we will mention it briefly as it relates to Noah and the Watchers. Most of us have been taught that God chose Noah to survive the flood because he was righteous. Noah was righteous, but God chooses him because his bloodline had not been tainted by the blood of the Watchers. We read in Genesis 6:8-9, *"But Noah found grace in the eyes of the Lord. These are the generations of Noah: Noah was a just man and perfect in his generations, and Noah walked with God."*

The phrase "perfect in his generations" refers to the bloodline of Noah. Five verses earlier, scripture mentioning the Watchers breeding into the human bloodline in Genesis 6:4, *"There were giants in the earth in those days; and also after that, when the sons of God came in unto the daughters of men, and they bare children to them, the same became mighty men which were of old, men of renown."*

The book of Enoch records it this way in chapter 10:

> [1.] Then said the Most High, the Holy and Great One spake, and sent Uriel to the son of Lamech, and said to him: [2.](Go to Noah) and tell him in my name "Hide thyself!" and reveal to him the end that is approaching: that the whole earth will be destroyed, and a deluge is about to come upon the whole earth, and will destroy all that is on it. [3.] And now instruct him that he may escape and his seed may be preserved for all the generations of the world.'

Scripture is clear that angels slept with human women. They and their offspring corrupted the world. God chose Noah, one of, if not the only bloodline still pure on the Earth to survive the coming deluge. The flood wiped out all flesh save those on the ark. The spirits of the Watchers were cast into the prison. The spirits of the Nephilim were cast into the valleys of the Earth.

Watchers in Scripture

In the book of Daniel, we read about an angel, the "prince of Persia" who fought against Michael, one of God's archangels, as well as the angel speaking. We read in Daniel 10:12-13, *"Then said he unto me, Fear not, Daniel: for from the first day that thou didst set thine heart to understand, and to chasten thyself before thy God, thy words were heard, and I am come for thy words. But the prince of the kingdom of Persia withstood me one and twenty days: but, lo, Michael, one of the chief princes, came to help me; and I remained there with the kings of Persia.*

In the book of Daniel, we see that some of God's angels are "princes" and are assigned oversight of the nations. In Daniel 12:1, Michael is the great prince who has charge of Israel, *"And at that time shall Michael stand up, the great prince which standeth for the children of thy people: and there shall be a time of trouble, such as never was since there was a nation even to that same time: and at that time thy people shall be delivered, every one that shall be found written in the book."*

Also in Daniel 10:20, we see that there is a prince (angel in charge) of Greece, *"Then said he, Knowest thou wherefore I come unto thee? and now will I return to fight with the prince of Persia: and when I am gone forth, lo, the prince of Grecia shall come."*

Indeed, today as Christians, we struggle against these princes. Many times, the New Testament uses the word "rulers" and "principalities" to describe these angels over the nations. We read in Ephesians 3:10, *"To the intent that now unto the principalities and powers in heavenly places might be known by the church the manifold wisdom of God."*

Our struggle culminates in the well-known scripture of Ephesians 6:12, *"For we wrestle not against flesh and blood, but against principalities, against powers, against the rulers of the darkness of this world, against spiritual wickedness in high places."*

In scripture, we have seen that angels chose to fall, to leave their spiritual realm and become mortal beings. There is evidence in scripture that some angels are made mortal not of their own will but of God's. We read about this in Psalm 82:7. The whole chapter is fascinating, so we'll look at Psalm 82 in its entirety:

> [1] God standeth in the congregation of the mighty;
> he judgeth among the gods.
> [2] How long will ye judge unjustly,
> and accept the persons of the wicked?
> *Selah.*
> [3] Defend the poor and fatherless:
> do justice to the afflicted and needy.
> [4] Deliver the poor and needy:
> rid them out of the hand of the wicked.
> [5] They know not, neither will they understand;
> they walk on in darkness:
> all the foundations of the earth are out of course.
> [6] I have said, Ye are gods;
> and all of you are children of the most High.
> [7] But ye shall die like men, and fall like one of the princes.
> [8] Arise, O God, judge the earth:
> for thou shalt inherit all nations.

There are so many fascinating concepts in this chapter. One concept that is relevant to our discussion about the princes in Daniel 10 is where Psalm 82 says the angels will "fall like one of the princes." This, I believe, is referring to the princes of Persia, Israel and Greece. God willing, my fourth book, after *Secret Societies*, will be a Marginal Mysteries book dedicated to angels.

The book of Enoch is not the only place where we read about the Watchers. They are mentioned by name in the Bible in Daniel 4:

> [13] I saw in the visions of my head upon my bed, and, behold, a watcher and an holy one came down from heaven;
>
> ...[17] This matter is by the decree of the watchers, and the demand by the word of the holy ones: to the intent that the living may know that the most High ruleth in the kingdom of men, and giveth it to whomsoever he will, and setteth up over it the basest of men.
>
> ...[23] And whereas the king saw a watcher and an holy one coming down from heaven, and saying, Hew the tree down, and destroy it; yet leave the stump of the roots thereof in the earth, even with a band of iron and brass, in the tender grass of the field; and let it be wet with the dew of heaven, and let his portion be with the beasts of the field, till seven times pass over him;

Indeed, I believe that "Satan" in Job 2 is not the Satan that we know as God's adversary but a Watcher. Satan means adversary or one who stands in the way of. The Hebrew for the adversary is *haś·śā·ṭān*. "Ha" is "the" in English. *Haś·śā·ṭān*, it's a title or office not a proper name (English: "the satan" is not "Satan").

There are instances in scripture where good spirits and people are called *haś·śā·ṭān* but the Masoretic does not translate them "Satan." King David is called *lə·śā·ṭān* (adversary) by the Philistines in 1 Samuel 29:4, *"And the princes of the Philistines were wroth with him; and the princes of the Philistines said unto him, Make this fellow return, that he may go again to his place which thou hast appointed him, and let him not go down with us to battle, lest in the battle he be an adversary to us: for wherewith should he reconcile himself unto his master? should it not be with the heads of these men?"*

The Angel of the Lord is called *lə·śā·ṭān* when he stands in the way of Balaam in Numbers 22:22, *"And God's anger was kindled because he went: and the angel of the Lord stood in the way for an adversary against him. Now he was riding upon his ass, and his two servants were with him."*

Keep these verses in mind as we read about this Watcher in Job 2:

> ¹ Again there was a day when the sons of God came to present themselves before the Lord, and Satan came also among them to present himself before the Lord. ² And the Lord said unto Satan, From whence comest thou? And Satan answered the Lord, and said, From going to and fro in the earth, and from walking up and down in it. ³ And the Lord said unto Satan, Hast thou considered my servant Job, that there is none like him in the earth, a perfect and an upright man, one that feareth God, and escheweth evil? and still he holdeth fast his integrity, although thou movedst me against him, to destroy him without cause. ⁴ And Satan answered the Lord, and said, Skin for skin, yea, all that a man hath will he give for his life. ⁵ But put forth thine hand now, and touch his bone and his flesh, and he will curse thee to thy face. ⁶ And the Lord said unto Satan, Behold, he is in thine hand; but save his life. ⁷ So went Satan forth from the presence of the Lord, and smote Job with sore boils from the sole of his foot unto his crown.

If "Satan" in Job 2 is indeed a Watcher, then it is the Watcher's job to accuse rich people on the Earth of loving their wealth more than God. Dr. Michael Heiser's words sum it up well:

> "He is, so to speak, Yahweh's eyes and ears on the ground, reporting what has been seen and heard… The function or office of the satan is why later Jewish writings began to adopt it as a proper name for the serpent figure from Genesis 3 who brought ruin to Eden… The dark figure of Genesis 3 was eventually thought of as the 'mother of all

adversaries,' and so the label satan got stuck to him. He deserves it. The point here is only that the Old Testament doesn't use that term for the divine criminal of Eden.

In Job 1 the satan and God converse about Job. The satan gets a bit uppity, challenging God about Job's integrity. We know the rest of the story—God gives the satan enough latitude to prove himself wrong, albeit at Job's expense."

For a more in-depth study on *haś·śā·ṭān* in Job 2, I recommend Dr. Michael Heiser's work on the topic, specifically in his book *The Unseen Realm*.

The Book of Enoch's Validity

The book of Enoch is generally grouped into 5 books:

The Book of Watchers (chs 1–36)

The Book of Parables (chs 37–71)

The Astronomical Book (chs 72–82)

The Animal Apocalypse or Book of Dreams (chs 83–90)

The Epistle of Enoch (chs 91–108)

Fragments of these sections have been found in the Dead Sea Scrolls with the exception of the Book of Parables. The fragments all date back to the centuries Before Christ – or year zero and back. The later scrolls that we have of the Book of Parables date to the 1st century A.D.

As we see, the Book of Enoch was written before the books of our New Testament, which were inspired by God. Could there be passages in the book of Enoch that mention celestial things about which mankind could not have known before God's inspired New Testament – before the words of Jesus even? If there are, then the book of Enoch becomes much more relevant in our exploration.

Fragments of the book of Enoch found in caves in Qumran. Credit: Dr. Avishai Teicher, Wikimedia

Many, if not all of the New Testament authors were scholars of the book of Enoch. In the Bible, Jude quotes Enoch almost verbatim. Enoch 1:9 reads, "And behold! He cometh with ten thousands of His holy ones To execute judgement upon all, And to destroy all the ungodly: And to convict all flesh Of all the works of their ungodliness which they have ungodly committed, And of all the hard things which ungodly sinners have spoken against Him."

Jude 14-15 reads, *"...Behold, the Lord cometh with ten thousands of his saints, To execute judgment upon all, and to convince all that are ungodly among them of all their ungodly deeds which they have ungodly committed, and of all their hard speeches which ungodly sinners have spoken against him."*

Does Jesus call the book of Enoch scripture? Jesus tells us in Mark 12:24-25 that the scriptures tell us that the angels in Heaven do not marry, *"And Jesus answering said unto them, Do ye not therefore err, because ye know not the scriptures, neither the power of God? For when they shall rise from the dead, they neither marry, nor are given in marriage; but are as the angels which are in heaven."*

Matthew 22:29-30 repeats the story, *"Jesus answered and said unto them, Ye do err, not knowing the scriptures, nor the power of God. For in the resurrection they neither marry, nor are given in marriage, but are as the angels of God in heaven."*

Yet, in our Old Testament, we don't find a verse saying that the angels in Heaven don't marry. We do however, find it in the book of Enoch chapter 15, *"³ Wherefore have ye left the high, holy, and eternal heaven, and lain with women, and defiled yourselves with the daughters of men and taken to yourselves wives, and done like the children of earth, and begotten giants (as your) sons?...⁶ But you wereformerly spiritual, living the eternal life, and immortal for all generations of the world. ⁷ And therefore I have not appointed wives for you; for as for the spiritual ones of the heaven, in heaven is their dwelling."*

Jesus was calling some unknown book, not included in our Old Testament, scripture. We can't be sure it was the book of Enoch. What we can be sure of, is that there was or is a book out there that the Messiah considered scripture. For the sake of space, here is a listing of nine times where Jesus appears to reference the book of Enoch:

> **John 14:2**, In my Father's house are many mansions...
> Enoch 45:3, In that day shall the Elect One sit upon a throne of glory, and shall choose their conditions and countless habitations.
>
> **John 12:36**, ...that ye may be the children of light.
> Enoch 108:11 {105: 25}, The good from the generation of light.
>
> **John 4:14**, ...but the water that I shall give him shall be in him a well of water springing up into everlasting life.
> Enoch 48:1 {48:1}, All the thirsty drank, and were filled with wisdom, having their habitation with the righteous, the elect, and the holy.

Matthew 5:5, Blessed are the meek, for they shall inherit the earth.

Enoch 5:7 {6:9}, The elect shall possess light, joy, and peace, and they shall inherit the earth.

John 5:22, For the Father judgeth no man, but hath committed all judgment unto the Son:

Enoch 69:27 {68:39}: The principal part of the judgment was assigned to him, the Son of man.

Luke 6:24, But woe unto you that are rich! for ye have received your consolation.

Enoch 94:8 {93:7}: Woe to you who are rich, for in your riches have you trusted; but from your riches, you shall be removed.

Matthew 19:28, ...That ye which have followed me, in the regeneration when the Son of man shall sit in the throne of his glory, ye also shall sit upon twelve thrones, judging the twelve tribes of Israel.

Enoch 108:12 {105:26}: And I will bring forth in shining light those who have loved My holy name, and I will seat each on the throne of his honour.

Matthew 26:24, ...but woe unto that man by whom the Son of man is betrayed! it had been good for that man if he had not been born.

Enoch 38:2 {38:2}: Where will the habitation of sinners be...who has rejected the Lord of spirits? It would have been better for them, had they never been born.

Luke 16:26, ...between us and you there is a great gulf fixed ...

Enoch 22: 9,11{22:10,12}, By a chasm...are their souls are separated.

What we've just read is the book of Enoch recording things <u>in</u> Heaven that no mortal should have known about. Yet, Jesus tells us about Heaven and confirms these things at least 300 years after Enoch was written.

I will speculate that a group of Jews known as the Masoretes, excluded the Book of Enoch from canon because the book helped confirm Jesus as the promised Messiah. This is the same group of Jews who translated the Masoretic text from which we get the King James Version of the Bible. These scribes were also heavily influenced by Augustine to remove the book of Enoch in the 4th century A.D.

To close out these thoughts and this chapter, I will repeat my belief that the book of Enoch is not inspired. I am giving my backup for why I believe it is correct history and can therefore help fill in the missing puzzle pieces of history.

CHAPTER 6

THE FLOOD

Because of the wickedness of man, the destruction of the Nephilim, and the corruption of the Watchers, God decided to reboot creation with the pure bloodline of Noah. We read in Genesis 6:13, *"And God said unto Noah, The end of all flesh is come before me; for the earth is filled with violence through them; and, behold, I will destroy them with the earth."*

The apocryphal book of Jasher records that God granted mankind a period of 120 years to repent and stop the destruction of the world with a great deluge. Noah and his grandfather Methuselah preach to mankind during this period. I don't believe that the book of Jasher is inspired by God, but we may get a puzzle piece or two if it fits in the incomplete biblical picture. We read in Jasher 5:7-8, *"Speak ye, and proclaim to the sons of men, saying, Thus saith the Lord, return from your evil ways and forsake your works, and the Lord will repent of the evil that he declared to do to you, so that it shall not come to pass. For thus saith the Lord, Behold I give you a period of one hundred and twenty years; if you will turn to me and forsake your evil ways, then will I also turn away from the evil which I told you, and it shall not exist, saith the Lord."*

This 120-year span is likely what inspired scripture is talking about in Genesis 6:3, *"And the Lord said, My spirit shall not always strive with man, for that he also is flesh: yet his days shall be an hundred and twenty years."*

From this passage, I believe the book of Jasher gains some relevance.

Genesis 7:11-12, *"In the six hundredth year of Noah's life, in the second month, the seventeenth day of the month, the same day were all the fountains of the great deep broken up, and the* *windows of heaven were opened. And the rain was upon the earth forty days and forty nights."*

"The windows of Heaven were opened" is either a fancy way to say that it rained, or it truly meant that a canopy was broken or melted. Some of the water that flooded the Earth came from the ice canopy above the skies.

The Meissner effect is the expulsion of a magnetic field from the interior of a material that is in the process of losing its resistance to the flow of electrical currents when cooled below a certain temperature. Very cold ice has magnetic properties therefore it's possible that much of the firmament would have been pulled to the north and south poles as it collapsed.

Scientists, in their attempt to disprove the Bible, will say that there was not enough water in the canopy to cover Mount Everest. We know that the flood covered the mountains from Genesis 7:18-20, *"And the waters prevailed, and were increased greatly upon the earth; and the ark went upon the face of the waters. And the waters prevailed exceedingly upon the earth; and all the high hills, that were under the whole heaven, were covered. Fifteen cubits upward did the waters prevail; and the mountains were covered."* and Psalm 104:6, *"Thou coveredst it with the deep as with a garment: the waters stood above the mountains."*

Not all of the rain would come from the ice canopy, however. Some of the water evaporated and re-rained down during the catastrophe as the rain cycle does today.

Scientists ignore the fact that not all of the water came from the ice canopy. Much of the flood water likely came from the fountains of the great deep – the bodies of water under our oceans and continents. As we just read in Genesis 7, the fountains of the great deep were broken up.

Scientists also ignore the possibility that Mount Everest was not yet formed at the time of the great flood of Noah. Most of our mountains today were likely formed in the years after the flood as the continents drifted and collided.

What are these "fountains of the great deep"? They are not the oceans, but bodies of water that exist under the crust and under the oceans even today.

In Sumerian and Akkadian mythology, Abzu is referred to as the primeval sea below the void space of the underworld and the Earth above.

A Fox News article published in 2019 states that scientists have found a gigantic freshwater aquifer hidden deep below the ocean. Discovery wrote a similar article in the same year stating that, "There's as much water inside the Earth as there is in all of the oceans." These fountains of the deep are 600 feet below the ocean floor.

You don't have to take scientists' word for it. Take God's. Psalm 136:6 reads, *"To him that stretched out the earth above the waters..."*

Psalm 24:1-2 reads, *"The earth is the Lord's, and the fulness thereof; the world, and they that dwell therein. For he hath founded it upon the seas, and established it upon the floods."*

Job 38:16 mentions the springs of the sea, *"Hast thou entered into the springs of the sea? or hast thou walked in the search of the depth?"*

It's great to see scientists "discover" things that the authors of the Old Testament knew 3,000 years ago. The breaking up of these fountains of the great deep is likely what separated the dry land we read about in Genesis 1:9, *"And God said, Let the waters under the heaven be gathered together unto one place, and let the dry land appear: and it was so."*

Have you ever noticed that the continents of the Earth look like a puzzle that has been taken apart? One German meteorologist named Alfred Wegener noticed and in 1915 the theory of plate tectonics was born. The theory describes the slow separation and movement of the continents.

We can see the scars of the breaking up of the fountains of the great deep today. The San Andreas fault is likely one of these locations. Earthquakes rip the ground open at an average of 2 miles per second. It would take less than four hours for one of these fault lines to circle the Earth.

After the breaking of the continents, they began to slowly drift apart. According to Genesis 10:25, it was in the days of Peleg that the Earth was divided, *"And unto Eber were born two sons: the name of one was Peleg; for in his days was the earth divided; and his brother's name was Joktan."*

The continents are still drifting today according to scientists. In fact, Forbes reports that 200 million years from now, the continents will collide in the Pacific forming a new continent called Amasia.

On March 22, 1998, a group of boys playing basketball outdoors in Monahans Texas narrowly missed being struck by a three-pound meteorite that crashed nearby. NASA scientists were baffled as to why the meteorite contained water and salt crystals.

The rock was likely shot into orbit when the fountains of the great deep broke up. The breaking up of the Earth's crust would have been violent enough to send rocks and water into outer space. This would also have helped with the breaking of the ice canopy.

Another possible cause for the breaking of the ice canopy would be a large comet. As we mentioned in the chapter on the firmament, the Earth's climate was likely springtime year-round and the Earth's axis, vertical. This comet could have sent the Earth into a wobble eventually settling on its current tilt of 23.5 degrees.

The Ark

According to Jasher 5:34-35, Noah and his sons made the ark in a five-year time period after which Noah took three granddaughters of Methuselah for wives for his sons, *"In his five hundred and ninety-fifth year Noah commenced to make the ark, and he made the ark in five years, as the Lord had commanded. Then Noah took the three daughters of Eliakim, son of Methuselah, for wives for his sons, as the Lord had commanded Noah."*

A common error among those who wish to disprove the Bible is that Noah couldn't possibly have gathered two of every animal. I agree that would be difficult, but Noah didn't gather them. We see in Genesis 6:19-20 that God brought the animals to the ark, *"And of every living thing of all flesh, two of every sort shalt thou bring into the ark, to keep them alive with thee; they shall be male and female. Of fowls after their kind, and of cattle after their kind, of every creeping thing of the earth after his kind, two of every sort shall come unto thee, to keep them alive."*

Some claim that Noah couldn't have fit all of the Earths animals on the ark. Noah easily fit two of every kind of animal on the ark. Noah even had room for 14 of every clean animal kind, according to Genesis 7:2-3, *"Of every clean beast thou shalt take to thee by sevens, the male and his female: and of beasts that are not clean by two, the male and his female. Of fowls also of the air by sevens, the male and the female; to keep seed alive upon the face of all the earth."*

It's unclear here whether Noah took seven or fourteen of every clean animal kind. Either way, there would still be room. First, one would not have to bring adults of each animal kind when yearlings would do. Why bring a fully grown T. rex when a snack-sized one will be able to repopulate the Earth? Young animals are smaller. They weigh less, eat less, sleep more, are tougher, and live longer allowing them to produce more offspring.

Also, a lot fewer animals were needed to repopulate the Earth than deniers like to claim. There are 35 species of Canidae (dogs, wolves, foxes, dingoes, etc.). Noah didn't need 70 different Canidae. He only needed two. After the deluge, just two could repopulate the Earth and new species would adapt thus changing their appearance. The same can be said for cats and indeed every animal kind on our planet. As a side note, we have a vast number of dog species because mankind has selectively bred dogs for various purposes.

What is meant by clean and unclean animals is not clear either. Levitical law informed the Israelites which animals were clean and unclean but wasn't given until 2,000 years later. Could clean and unclean mean the animals that were and were not corrupted by the Nephilim in Genesis 6:12? *"And God looked upon the earth, and, behold, it was corrupt; for all flesh had corrupted his way upon the earth."*

As we looked at in the chapter on the Watchers, we also read about the giants sinning against the animals in Enoch 7:

> [4.] the giants turned against them and devoured mankind.
> [5.] And they began to sin against birds, and beasts, and reptiles, and fish, and to devour one another's flesh, and drink the blood. [6.] Then the earth laid accusation against the lawless ones.

We read the ark's dimensions in Genesis 6:14-15, *"Make thee an ark of gopher wood; rooms shalt thou make in the ark, and shalt pitch it within and without with pitch. And this is the fashion which thou shalt make it of: The length of the ark shall be three hundred cubits, the breadth of it fifty cubits, and the height of it thirty cubits."*

The ark of Noah would have been able to house about 2,150,000 sheep. We don't know how many species of animals have gone extinct over the last 5,000 years. Estimating all of the known animal kinds, bringing two of each, a maximum of 50,000 sheep-sized animals would have needed to be brought on board. That's 1/43rd of the ark's capacity. If Noah had taken fourteen of every animal kind, a maximum of 350,000 animals would have been on board. That's 1/6th of the ark's capacity.

Rainbow

After the flood, Elohim promises to never destroy the Earth with water again. We read about the rainbow in Genesis 9:

> [11] And I will establish my covenant with you, neither shall all flesh be cut off any more by the waters of a flood; neither shall there any more be a flood to destroy the earth. [12] And God said, This is the token of the covenant which I make between me and you and every living creature that is with you, for perpetual generations: [13] I do set my bow in the cloud, and it shall be for a token of a covenant between me and the earth. [14] And it shall come to pass, when I bring a cloud over the earth, that the bow shall be seen in the cloud: [15] And I will remember my covenant, which is between me and you and every living creature of all flesh; and the waters shall no more become a flood to destroy all flesh. [16] And the bow shall be in the cloud; and I will look upon it, that I may remember the everlasting covenant between God and every living creature of all flesh that is upon the earth. [17] And God said unto Noah, This is the token of the covenant, which I have established between me and all flesh that is upon the earth.

I speculate that when one looks at the rainbow, one is truly looking at something from the spiritual realm. I don't believe that the rainbow was created on one of the six days of creation but that it was already in the spiritual realm and revealed to mankind as a promise. We read about the emerald rainbow in Heaven in Revelation 4:1-3, *"After this I looked, and, behold,* a door was opened in heaven: and the first voice which I heard was as it were of a trumpet talking with me; which said, Come up hither, and I will shew thee things which must be hereafter. And immediately I was in the spirit: and, behold, a throne was set in heaven, and one sat on the throne. And he that sat was to look upon like a jasper and a sardine stone: and there was a rainbow round about the throne, in sight like unto an emerald."

We read about a mighty angel with a rainbow on his head in Revelation 10:1, *"And I saw another mighty angel come down from heaven, clothed with a cloud: and a rainbow was upon his head, and his face was as it were the sun, and his feet as pillars of fire."*

I believe that the rainbow is something in the spiritual realm that God allows us to see after his promise to Noah and mankind. If you've ever wanted to see something alien, then think about this the next time you see a rainbow.

There are many times in scripture where the veil between the spiritual and physical realms is parted. Saul asks the witch of Endor, at night, to bring up Samuel, who was dead, in a strange and telling passage. 1 Samuel 28 reads:

> [8] And Saul disguised himself, and put on other raiment, and he went, and two men with him, and they came to the woman by night: and he said, I pray thee, divine unto me by the familiar spirit, and bring me him up, whom I shall name unto thee....[11] Then said the woman, Whom shall I bring up unto thee? And he said, Bring me up Samuel....[13] And the king said unto her, Be not afraid: for what sawest thou? And the woman said unto Saul, I saw gods ascending out of the earth. [14] And he said unto her, What form is he of? And she said, An old man cometh up; and he is covered with a mantle. And Saul perceived that it was Samuel, and he stooped with his face to the ground, and bowed himself.

This is a telling passage indeed that is worthy of study because God chose to include it in His word. What were the "gods ascending out of the earth" before Samuel came "up"? Were they angels escorting Samuel from wherever he came up? Were the escorting angels necessary in light of the great gulf mentioned in Luke 16:25-26, *"But Abraham said, Son, remember that thou in thy lifetime receivedst thy good things, and likewise Lazarus evil things: but now he is comforted, and thou art tormented. And beside all this, between us and you there is a great gulf fixed: so that they which would pass from hence to you cannot; neither can they pass to us, that would come from thence."*

Did God make an exception for Saul's request because he was God's anointed? Also, this does not appear to be a demon but Samuel himself. Saul also asks what form Samuel took before seeing for himself as if he's done it before.

In 2 Kings 6 the Lord opens the eyes of Elisha's young servant and allowed him to see the spiritual realm:

> 15 And when the servant of the man of God was risen early, and gone forth, behold, an host compassed the city both with horses and chariots. And his servant said unto him, Alas, my master! how shall we do? 16 And he answered, Fear not: for they that be with us are more than they that be with them. 17 And Elisha prayed, and said, Lord, I pray thee, open his eyes, that he may see. And the Lord opened the eyes of the young man; and he saw: and, behold, the mountain was full of horses and chariots of fire round about Elisha.

Hebrews 13:2 says that some people meet angels without knowing it, *"Be not forgetful to entertain strangers: for thereby some have entertained angels unawares."*

In Numbers 22, Balaam's donkey sees an angel when Balaam does not:

> 22 And God's anger was kindled because he went: and the

angel of the Lord stood in the way for an adversary against him. Now he was riding upon his ass, and his two servants were with him. 23 And the ass saw the angel of the Lord standing in the way, and his sword drawn in his hand: and the ass turned aside out of the way, and went into the field: and Balaam smote the ass, to turn her into the way. ...31Then the Lord opened the eyes of Balaam, and he saw the angel of the Lord standing in the way, and his sword drawn in his hand: and he bowed down his head, and fell flat on his face.

That's a fascinating rabbit trail about our interaction with the spiritual realm. Now, I will get back to the flood.

Extra-Biblical Flood Accounts

There exist flood legends from over 270 cultures from all around the world. In thirty-two of those accounts, a favored family was saved. In twenty-one of those accounts, survival was due to a boat. In nine of those accounts, animals are saved.

Why are there so many stories of a global flood? As God says in Genesis 10:32, all nations and races come from the eight survivors of the great flood, *"These are the families of the sons of Noah, after their generations, in their nations: and by these were the nations divided in the earth after the flood."*

Since this is true, no matter the color of skin or location on Earth, every culture, that recorded history, should tell of similar events. That's exactly what we find.

After the confusion at the Tower of Babel, for many cultures, written characters were developed to represent each of their new languages. The American Lutheran theologian John Warwick Montgomery's words sum it up well:

> "The destruction of well nigh the whole human race, in an early age of the world's history, by a great deluge, appears to have so impressed the minds of the few

survivors, and seems to have been handed down to their children, in consequence, with such terror-struck impressiveness, that their remote descendants of the present day have not even yet forgotten it. It appears in almost every mythology, and lives in the most distant countries, and among the most barbarous tribes."

Our most ancient accounts of the flood are:

Source	Culture	Est. Date	Surviving Man	God
Bible	Hebrew	1445 BC	Noah	Elohim
Eridu Genesis	Sumerian	2300 BC	Ziudsura	Enlil & An
The Atrahasis	Akkadian	1700 BC	Atrahasis	Enlil
Epic of Gilgamesh	Sumerian	2150 BC	Utnapishtam	Anu
Book of the Heavenly Cow	Egyptian	2181 BC		Ra
Book of Documents	Chinese	1000 BC	Nuwa	
Plato & Homer	Greek	360 BC	Deucalion	Zeus

Chinese texts describe a gigantic flood so huge that it rose to Heaven and drowned all the mountains and living things. Most of these texts are found in the ancient Book of Documents (Shu Jing). Though the original texts were destroyed by emperor Qin Shi Huang around 200 B.C., most of the remaining texts were re-compiled by historians Confucius and Si Ma Qian.

Most of the ancient Chinese writing characters were pictographic. Today only about 4 percent remain so. Since most of the ancient texts were destroyed, scholars must turn to deciphering the pictographic characters to interpret their meanings and the stories they tell.

On ancient Chinese bronzeware and oracles bones we find a particularly interesting story. The ancient Chinese symbol for boat is comprised of three symbols: vessel, eight, mouths (persons).

船 = 舟 口 几

VESSEL EIGHT MOUTHS

Could this be referring to anything other than Noah and the survivors of the great flood? These Chinese carvings support the idea of the fountains of the great deep. They record the violent bursting out of water from the pit or abyss upward into the sky. The flood waters rose up to the sun.

VIOLENT FLOOD = **WATER STREAM** **TWO HANDS** **EJECTING OUT** **SUN**

The texts also tell of a boat to be built as a refuge during a judgment.

BUILD OR CREATE = **BOAT** **ROOF** **JUDGEMENT** **TRAVEL**

The carvings also show characters that translate as a ship stopping on a mountain with pairs of animals. The knife symbolizes a form of punishment.

肖 . 身 . 𩪠 = 舟 ⺋ ⺂ 刀

BOAT STOP MOUNTAIN KNIFE

𦫿 = 𣎆 屾

PAIR OF ANIMAL HEADS **SHIP**

There are many characters in this carving that shows animals just after the boat comes to rest on the mountain.

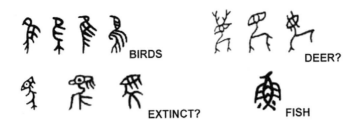

BIRDS

DEER?

EXTINCT?

FISH

From these characters, we see the story of judgment which resulted in a ship filled with creatures and people coming to rest on a mountain. The study is fascinating.

Taoism is an ancient Chinese philosophy and religion. According to Taoist stories, eight immortals crossed the sea on a ship. In a story at least as old as 2205 B.C., Nuwa was a man who turned back a catastrophic flood that covered the whole world. In the story, the heavens are broken and the nine states of China experienced continental shift and were split. The water covered the mountains and all living things perished.

The Greeks have several versions of a flood myth. They all include king Deucalion and his wife Prrha who escaped from a great flood by floating in a chest that lands on a mountain.

A Hindu flood myth from the 6th century B.C. tells of a hero, Manu, who was advised by a fish to build a boat in order to escape the coming flood. During the flood, the fish towed the ship to a mountaintop.

In the Norse flood story, the gods are at war amongst themselves. When Odin and his brothers Villi and Ve killed the giant Ymir, the blood that poured forth flooded the Earth. A frost giant named Bergelmir and his wife made an ark and were saved to repopulate the Earth.

In the Aztec flood legend, Nata and Nena are warned by the god Titlacauan that a great flood was coming. Nata and Nena hollowed out a cypress tree and used it as a boat to survive the flood.

The Atrahasis, as we looked at in the chapter on creation, is the Akkadian/Babylonian epic written around 1700 B.C. First, the gods created mankind to work on the Earth to labor for them. Eventually, Enlil the king of the gods, was unhappy with how loud humans were so he planned to send a flood to destroy them. Enki told a righteous man, Atrahasis, how to survive the flood by building an ark. Enki instructs Atrahasis to seal two of every kind of animal within the ark. We read about the deluge on Tablet III, Dalley 31:

> The flood came out...No one could see anyone else
> They could not be recognized in the catastrophe
> The Flood roaraed like a bull
> Like a wild ass screaming, the winds howled
> The darkness was total, there was no sun.

After the flood, the gods curse mankind by sending evil spirits to keep their population down. We will read in the chapter on the Nephilim about the evil spirits that appear on the Earth after the flood of Noah. The gods also lower the lifespans of mankind.

The Eridu Genesis is the Sumerian Flood Story and was written around 2300 B.C. The part of the tablet that tell why the gods sent a great flood is missing. Ziudsura is the good man that Enki chooses to survive the flood. Ziudsura builds a large boat and fills it with animals. Fragments of the Eridu Genesis tablet are missing so the text is broken up:

> I will speak words to you; take heed of my words, pay attention to my instructions. A flood will sweep over the
> ... A decision that the seed of mankind is to be destroyed has been made. The verdict, the word of the divine assembly, cannot be revoked. The order announced by An and Enlil cannot be overturned. Their kingship, their term has been cut off; their heart should be rested about this...Here there are approximately 38 lines missing

All the windstorms and gales rose together and the flood swept over the land. After the flood had swept over the land, and waves and windstorms had rocked the huge boat for seven days and seven nights, Utu the sun-god came out, illuminating heaven and earth.

The Epic of Gilgamesh, though recorded around 2000 B.C, is the oldest known flood story (it was passed orally before the Eridu Genesis). The epic is recorded on twelve stone tablets which are among the first pieces of literature in history. In the epic, the flood has already happened but Gilgamesh searches for Utnapishtam, the survivor of the flood, to find immortality. Giglamesh discovers that mortals cannot have immortality unless granted by the gods.

A scene from the Book of the Heavenly Cow as depicted in the tomb of Seti I

The Egyptian tale of the flood is found in *The Book of the Heavenly Cow* written as early as 2181 B.C. In the story, Ra created humans. Humans rebelled against the gods so Ra has Hathor destroy humanity. Hathor does so partially by drowning the people with beer. We read about the flood in *The Book of the Heavenly Cow*:

The Majesty of the King of Upper and Lower Egypt, Ra, got up early in the deep of the night in order to have this intoxicating draught poured out. Then the fields became filled to a height of three palms with the liquid through the power of the Majesty of this god. This goddess set out in the morning and she found these fields inundated. Her face became delighted thereat.

After the destruction, Ra leaves the Earth in the hands of his lesser gods. We will read about the lesser gods being given dominion of the Earth in the chapter on the postdiluvian Earth.

The Enchanted Mountain is a spur of the Blue Ridge mountains binding Tennessee and North Carolina. One hundred thirty-six animal foot impressions have been found in the rocks of this area known as Track Rock Gap. Human feet, including a giant 17-inch six-toed footprint, are as well impressed into the rocks. These tracks are said to be 4,000 years old.

The Indian tradition asserts the world was once inundated by a great flood. Everything was destroyed save one family, together with the various animals necessary to replenish the Earth. They say that the great canoe once rested on this spot and the footprints in the stone are those of the animals as they disembarked the great canoe.

As we can see, the event of the global flood is almost universally recorded by the ancients, yet modern-day scholars reject the idea. Again, they do so because they will lose their jobs if they even suggest that the great flood actually took place. It's not the agenda of the elites to allow for God's existence, much less His wrath. This coming from the same science that can't tell you what a woman is!

Evidence of The Flood

The question is not whether the Earth has cracks, the question is how did they get there? As mentioned earlier, we can see the scars left when the fountains of the deep broke apart the continents. These scars are most noticeable at the world's tectonic plates.

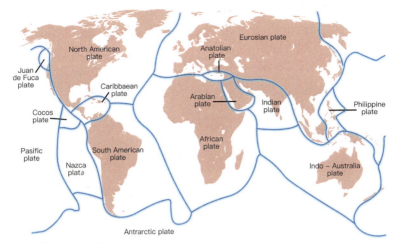

The Earth's tectonic plates

If you've ever heard someone, say that California will fall into the ocean one day as the result of a massive earthquake, that is not fully correct. Only a small portion of the state (the whole coast) would fall off as seen above. You can see the scars of the San Andreas fault:

The San Andreas Fault in California, USA.

Not only do we see the scars of the breaking up of Pangea but we can also see the massive amount of water runoff in satellite imagery today. It's clear through images that the northern half of the continent of Africa, where the Sahara Desert is today, was underwater in the past. We studied this in our chapter on Atlantis in my previous book, *Ancient Cities and the gods Who Built Them*.

Notice not only the flow lines moving westward across Africa, but the silt buildup in the Atlantic Ocean where sand was pulled off of the continent and deposited into the ocean. The dark spots are exposed bedrock where the flowing water stripped off all of the soil.

Water striations in northwest Africa | Credit: Google Earth

Water runoff cutting through a previous lava flow at Emi Koussi in Africa | Public Domain

Deniers will argue that this wasn't due to water, but due to wind. The white spots in the desert are actually salt, remnants of the salt water ocean that once flowed on top. Also, the 10 windiest places on Earth, such as Mount Everest, don't show this same soil movement. You can also see this water runoff cut through a 12,000-year-old lava flow, proving that this flood event was relatively recent.

We see this flow scarring all over the globe. It's obvious in Australia in the region of Lake Eyre. The Gulf of Mexico is full of silt where it borders the southern United States. This silt came from the North American continent as far up as Canada as the flood waters receded and pulled the silt into the gulf. Water runoff can also be seen in Washington state. Striations were left behind as the water flowed from the Rocky Mountains in the east to the Pacific Ocean to the west.

Whale bones in the whales valley in Egypt., Credit: AhmedMosaad, Wikipedia

Scientists have found hundreds of sea creature remains in places that are nowhere near oceans. One of these places is Wadi al Hitan in Egypt. It is a paleontological site that was designated a UNESCO World Heritage Site in July of 2005 for its hundreds of fossils.

In 2000, the scientific community was astonished by the discovery of a stone age settlement at the bottom of the Black Sea, 12 miles off the coast of northern Turkey. Stone tools and wooden beams are among the well-preserved remnants of the structure.

The discovery supports the theory that the seabed was once populated by a farming community who fled a flood. Terry Garcia, head of mission programmes for National Geographic said of the discovery:

> The significance of this find is that for the first time we will have established that human beings had settled in

this area and were occupying this area at the time of this cataclysmic event.

Fossils of giant ichthyosaurs were found at three different sites in the eastern Switzerland alps. In 2022, one of the fossils was found at 8,990 feet above sea level. While it is possible for ancient sea beds to end up on mountains through plate tectonics, evolutionists will not admit that ancient sea creatures may have been trapped in the mountains during receding waters.

Over 800 human skeletons were found at Roopkund Lake., Credit: Schwiki Wikipedia

One mystery that still baffles scientists today is that of Roopkund Lake high up in the Himalayan mountains. The lake is small and a five-day hike from the nearest village. What's fascinating is that over 800 human skeletons have been found at Roopkund Lake.

How did these skeletons get here? National Geographic formulated a theory that they were pilgrims that encountered a hailstorm that killed them. Others speculate that thags (Hindi bandits from whom we get the English word thug) murdered these people and dumped their bodies there. Both of these theories are possible.

I'll theorize that they were antediluvian people attempting to flee Noah's flood. As the waters rose, they trekked higher and higher into the tallest mountains around until the flood overtook them.

Radiocarbon dating has been spotty on the remains, not to mention the flaws in radiocarbon dating itself. The flood being the demise of these 800 people is an interesting theory. We may never know.

A crinoid fossil in a boulder at 8,300 foot elevation., Credit: tidalwaveyahweh TikTok

We also find sea creatures on the tops of mountains. Reported by Newsweek in 2021, TikTok user @tidalwaveyahweh shared a video of a fossilized crinoid (related to star fish) in a massive boulder at an elevation of 8,300-feet.

When clams die, their muscles relax, thus opening their shells. All over the world, we have clams that are fossilized in their closed state. This means that these clams were buried while still alive – caused by sudden water and mud movement.

Not only do we have hundreds of skeletons and fossilized sea life on the Earth's mountaintops, but we may have discovered the landing site of Noah's ark. The Bible says that Noah's ark came to rest upon the mountains of Ararat in Genesis 8:4, "*And the ark rested in the seventh month, on the seventeenth day of the month, upon the mountains of Ararat.*"

The Ararat mountains are in eastern Turkey today. While the ark likely didn't come to rest on what we know as Mount Ararat, it did come to rest in the mountains.

For me, the most likely location of the ark's resting place is that of Durupinar, Turkey. Durupinar is a site 17 miles south of Mount Ararat, still in the mountains of Ararat. At the site is a 515-foot-long (300 Egyptian cubits) boat shaped impression in a mud-flow. The site is officially recognized as the landing of Noah's ark by the Turkish government. The site was discovered in 1948 by a shepherd from nearby Uzengili village.

The Durupinar Site showing the ark imprint and Gilgamesh's Mashu mountain., Credit: Wikkiwooki, Wikipedia

Eleven years later, Turkish Air Force Captain Ilhan Durupinar identified the formation in an aerial photo taken during a NATO mapping mission. The boat shaped impression can easily be seen in the photo. The impression is not a naturally occurring one according to radar scans. At the site archaeologists have found petrified wood. Both the vertical ribs and horizontal beams are symmetrical indicating a structure.

Three metal rivets were found at the ark site made of alloys. Noah likely knew about metallurgy as he was a few generations after Tubalcain who worked with alloys. Genesis 4 tells us that brass and iron were worked by Tubalcain 22, "And Zillah, she also bare Tubalcain, an instructer of every artificer in brass and iron: and the sister of Tubalcain was Naamah."

Left: 1959 Durupinar aerial photo | Right: David Fasold with a drogue stone., Credit: Robert Michelson, Wikipedia

Drogues are stone weights used to stabilize vessels in ancient times. Over 20 massive anchor stones have been found in the area. Pictured with one of these stones is Dave Fasold who wrote the book, The Ark of Noah. The crosses were likely carved into the stones thousands of years later.

Also at the site are the ruins of the biblical city of Mesha known as the first city built after the flood. We read about the city in Genesis 10:30 as Shems (2nd son of Noah) dwelling, *"And their dwelling was from Mesha, as thou goest unto Sephar a mount of the east."*

Mesha means "to be drawn out of water," a name most likely given to it by Noah. Mesha is a variation of Moshe meaning, "saved through water," as Moses (Moshe in Hebrew) was saved from the Nile.

Mashu is also mentioned in the Epic of Gilgamesh as the mountains to which Gilgamesh traveled to find Utnapishtim (Noah), "Mashu the name of the hills; as he reach'd the Mountains of Mashu, The name Gilgamesh means "the man who revealed Mesha."

CENOZOIC	HOLOCENE	10,000 Years
	PLEISTOCENE	1.8 Millions Years ago
	PLIOCENE	5.3 Millions Years ago
	MIOCENE	23 Millions Years ago
	OLIGOCENE	33.9 Millions Years ago
	EOCENE	55.8 Millions Years ago
	PALEOCENE	65.5 Millions Years ago
MESOZOIC	CRETACEOUS	145.5 Millions Years ago
	JURASSIC	199.6 Millions Years ago
	TRIASSIC	251.2 Millions Years ago
PALEOZOIC	PERMIAN	299 Millions Years ago
	PENNSYLVANIAN	318 Millions Years ago
	MISSISSIPPIAN	359.2 Millions Years ago
	DEVONIAN	416 Millions Years ago
	SILURIAN	443 Millions Years ago
	ORDOVICIAN	488.3 Millions Years ago
	CAMBRIAN	542 Millions Years ago
	PROTEROZOIC	2.5 Billion Years ago
	ARCHEAN	

The geologic column showing the different layers of strata in the earth.

Because of time and natural events such as flooding, layers of dirt and debris stack up and build what we know as strata. As we dig deeper into the Earth, we uncover layers of older ground.

In 1938, Frank and Emma Hahn, while out hiking in London Texas, discovered a rock outcropping that contained a hammer. The hammer dated to only 350 years old, but the strata dated to 135 million years ago.

In 1944 Newton Anderson, a 10-year-old boy, found a brass bell inside of a lump of coal. The coal was mined in Upshur County West Virginia. Coal is said to be hundreds of millions of years old. Analyzed by the University of Oklahoma, the bell was found to be composed of a strange and unknown mixture of metals. The bell looks like a normal bell with a winged creature on top. The creature looks suspiciously like the Babylonian wind demon named Pazuzu.

Evolutionists say that over billions of years, animal species that have since gone extinct are found at layers lower than that of humans. What would it do to their theory if we were to find evidence of humans with dinosaurs?

After a 1920s flood in Glen Rose Texas, thousands of dinosaur footprints were revealed, preserved in rock. You can see these prints today at Dinosaur Valley State Park. In 1930 Roland Bird, a field explorer for the American Field Museum reported giant 15- to 20-inch-long human footprints with the dinosaur tracks. Dinosaur and human tracks are also found together embedded in rock in Holyoke, Mass.

Left: The London Hammer | Right: Paluxy River tracks in Glen Rose, Tex. Credit: S.J. Miba, Wikimedia

In 1866, workers found a human skull buried 130 feet below the surface below a layer of lava in Calaveras County, Calif. Geologists concluded that the skull was from the Pliocene era (5 million years ago). The skull is in Peabody Museum in Cambridge, Mass. today.

Where I live in Gray, Tenn., in 2000, road workers uncovered a forested pond ecosystem said to be over 4 million years old. Paleontologists discovered over 200 species of ancient animals and plants including mammoths, wooly rhinoceroses, and the only completely known fossil of a red panda in the Americas.

Needless to say, these species no longer populate the Appalachian Mountains of Tennessee. Whichever male and female of these species survived Noah's flood, didn't see fit to head back to Tennessee. Hundreds of times, I drove just meters above these amazing fossils without even knowing it.

Since East Tennessee State University is in charge of the site, they are responsible for its funding. The fossil site was in my State Representative district in Washington County, Tennessee. One year, folks from the university asked that I support $5 million in state funding for the site. I told them that if they put a plaque out front saying that these remains are from the antediluvian Earth created by God some 7,500 years ago, then I would be happy to vote for it.

An exhibit at the Gray Fossil Site & Museum in Gray, Tenn. Credit: Zilantophis, Wikipedia

Evolutionists will say that the geologic column is evidence for evolution. I say that the geologic column is evidence for a global flood. With a global flood, the water bottom will be churning violently burying all kinds of things in chaotic order. You will also find evidence of sea creatures in the mountains and in the deserts.

As we discussed in the chapter on the firmament, polystrate fossils are fossilized trees that extend through multiple layers of strata. Trees that stand through multiple layers of stratum are evidence against evolution. Trees don't just stand upright for billions of years while layers of strata build up around them, eventually burying them.

2 Peter 3:5, *"For this they willingly are ignorant of, that by the word of God the heavens were of old, and the earth standing out of the water and in the water."*

So many today are willingly ignorant of the clear evidence of the great flood of Noah. This event would change creation forever.

CHAPTER 7

THE NEPHILIM

Have you ever been responsible for someone or recommended someone to an employer? What if the Watchers didn't mean to corrupt the Earth so totally? What if they thought that Elohim could forgive them for mingling with women and teaching forbidden knowledge? What if they then began to realize that their offspring were utterly corrupting the Earth and finally realized how much trouble they were in? Now, we'll take a closer look at the Nephilim, the offspring of the Watchers.

Though the Nephilim mass killed each other before the flood, they did come back afterward. How they came back, I'm not sure. One possibility is that the daughters-in-law of Noah had giant's blood in their veins. Another possibility is that more angels rebelled and slept with women after the flood. Whatever the reason, God says they were on the Earth after the flood in Genesis 6:4, *"There were giants in the earth in those days; and also after that, when the sons of God came in unto the daughters of men, and they bare children to them, the same became mighty men which were of old, men of renown."*

What we will see in this chapter is that the giants lived in the time of Moses and through to the time of David. The second appearance of the Nephilim seems much more limited than the first appearance. This is due to Israel, God's chosen people. God uses various kings to wipe out the Nephilim particularly Joshua and David. Moses' army kills Og in Deuteronomy 3:

> *¹Then we turned, and went up the way to Bashan: and Og the king of Bashan came out against us, he and all his people, to battle at Edrei....³ So the Lord our God delivered into our hands Og also, the king of Bashan, and all his people: and we smote him until none was left to him remaining....¹¹ For only Og king of Bashan remained of the*

remnant of giants; behold his bedstead was a bedstead of iron; is it not in Rabbath of the children of Ammon? nine cubits was the length thereof, and four cubits the breadth of it, after the cubit of a man.

The Hebrew word for "giants" in verse 11 is *rᵊpā'îm* (Repha'im) which is only one race of giants. Moses' army only killed off the Repha'im and not all of the races of giants.

In this chapter, we will also see that the Nephilim, though not as powerful as their forefathers, were kings that ruled over the postdiluvian humans. They were worshiped as gods and demi-gods. Enoch gives credence to this in chapter 19:1-2, *"And Uriel said to me: 'Here shall stand the angels who have connected themselves with women, and their spirits assuming many different forms are defiling mankind and shall lead them astray into sacrificing to demons as gods, (here shall they stand,) till the day of the great judgement in which they shall be judged till they are made an end of. And the women also of the angels who went astray shall become sirens.'"*

It seems here that the female Nephilim play a role in history, and obviously, mythology. Sirens appear in the Greek epic the Odyssey as half bird-half women who sexually lure sailors to their deaths. This isn't far off from the succubus, Lilith in Sumerian, Roman, Greek and Egyptian legends who is regarded as the mother of all succubi.

Giants in Scripture

The giants are mentioned dozens of times in scripture. They are mentioned even more if you believe Enoch that their spirits are the evil spirits that dwell on the Earth. We will study about the Nephilim becoming the evil spirits in a few pages. There are many races of giants in the Bible including Anakim, Rephaim, Emims, Zamzummims (Zuzim) and Amorites. We already read about the Rephaim whose king was Og of Bashan. Deuteronomy 2 tells us of the Emims, Anakims and Zamzummims: [10] The

Emims dwelt therein in times past, a people great, and many, and tall, as the Anakims; [11] Which also were accounted giants, as the Anakims; but the Moabites called them Emims...[20] (That also was accounted a land of giants: giants dwelt therein in old time; and the Ammonites call them Zamzummims;

Amos 2:9 Tells us of the Amorites, *"Yet destroyed I the Amorite before them, whose height was like the height of the cedars, and he was strong as the oaks; yet I destroyed his fruit from above, and his roots from beneath."*

When God sends Joshua on his conquests, sometimes He tells him to just conquer the armies. Sometimes God tells him to wipe them out, including the women, children and livestock. I always wondered why God told Joshua to do this. After learning about what the giants had done to Elohim's creation in the antediluvian world, I now understand. Wherever God tells Joshua to utterly wipe them out, they were a race of giants.

Eventually, Joshua completes God's task. There are still Anakim left in a few places, however. We read in Joshua 11:21-22, *"And at that time came Joshua, and cut off the Anakims from the mountains, from Hebron, from Debir, from Anab, and from all the mountains of Judah, and from all the mountains of Israel: Joshua destroyed them utterly with their cities. There was none of the Anakims left in the land of the children of Israel: only in Gaza, in Gath, and in Ashdod, there remained."*

Goliath is the most renowned giant in scripture. Notice in scripture that Goliath was from Gath – one of the places in which Joshua left the giants alive 400 years earlier. We read in 1 Samuel 17:4-7:

And there went out a champion out of the camp of the Philistines, named Goliath, of Gath, whose height was six cubits and a span. And he had an helmet of brass upon his head, and he was armed with a coat of mail; and the weight of the coat was five thousand shekels of brass.

And he had greaves of brass upon his legs, and a target of brass between his shoulders. And the staff of his spear was like a weaver's beam; and his spear's head weighed six hundred shekels of iron: and one bearing a shield went before him.

There is much speculation about whether Goliath was a giant in this passage. Though the Masoretic text records him as being approximately 9 feet 9 inches tall, the Greek Septuagint and Dead Sea Scrolls record him as being 6 feet 9 inches tall. The fact that his gear was so heavy tells us that he was of enormous size and strength. Also, the Israelite warriors were afraid of Goliath in 1 Samuel 17:11, *"When Saul and all Israel heard those words of the Philistine, they were dismayed, and greatly afraid."*

2 Samuel 21 confirms Goliath as being a giant from Gath:

> [15] Moreover the Philistines had yet war again with Israel; and David went down, and his servants with him, and fought against the Philistines: and David waxed faint. [16] And Ishbibenob, which was of the sons of the giant, the weight of whose spear weighed three hundred shekels of brass in weight, he being girded with a new sword, thought to have slain David. [17] But Abishai the son of Zeruiah succoured him, and smote the Philistine, and killed him. Then the men of David sware unto him, saying, Thou shalt go no more out with us to battle, that thou quench not the light of Israel. [18] And it came to pass after this, that there was again a battle with the Philistines at Gob: then Sibbechai the Hushathite slew Saph, which was of the sons of the giant. [19] And there was again a battle in Gob with the Philistines, where Elhanan the son of Jaareoregim, a Bethlehemite, slew the brother of Goliath the Gittite, the staff of whose spear was like a weaver's beam. [20] And there was yet a battle in Gath, where was a man of great stature, that

had on every hand six fingers, and on every foot six toes, four and twenty in number; and he also was born to the giant. ²¹ And when he defied Israel, Jonathan the son of Shimeah the brother of David slew him. ²² These four were born to the giant in Gath, and fell by the hand of David, and by the hand of his servants.

We'll talk about the 6 fingers in a little bit. Most stories depict David as a shepherd boy but before he fought against Goliath, he was a "mighty valiant man, and a man of war" who had already killed a bear and a lion. We read in 1 Samuel 16:18, *"Then answered one of the servants, and said, Behold, I have seen a son of Jesse the Bethlehemite, that is cunning in playing, and a mighty valiant man, and a man of war, and prudent in matters, and a comely person, and the Lord is with him."*

One of David's might men, Benaiah the son of Jehoiada, defeated a large Egyptian man measuring 7 feet 6 inches in 1 Chronicles 11:23, *"And he slew an Egyptian, a man of great stature, five cubits high; and in the Egyptian's hand was a spear like a weaver's beam; and he went down to him with a staff, and plucked the spear out of the Egyptian's hand, and slew him with his own spear."*

Among King David's mighty men were two lion-faced men. We read about them in 1 Chronicles 12:8, *"And of the Gadites there separated themselves unto David into the hold to the wilderness men of might, and men of war fit for the battle, that could handle shield and buckler, whose faces were like the faces of lions, and were as swift as the roes upon the mountains."*

The idea of men who had faces like lions is fascinating. Scripture is not the only place where we find men described like this. An Egyptian text called The Craft of the Scribe, dating to 1250 BC, discusses a Canaanite mountain pass: The face of the pass is dangerous with Shasu, hidden under the bushes. Some of them are 4 or 5 cubits, nose to foot, with wild faces.

Not only did these Shasu have wild faces, but they were also around 8 feet tall! They lived in the mountains and the lion-faced men lived in the wilderness. Could this Egyptian text be referring to the same men mentioned in scripture?

After this point in scripture, giants are no longer mentioned with the exception of a retrospective mention in the book of Amos which we read earlier.

How big were the giants in scripture? Goliath was 9 feet 9 inches according to the Masoretic. King Og's bed of iron (whether sleeping bed or coffin, it doesn't matter) was 13 feet 6 inches. Since most creatures in the fossil record appear to have been twice the size of what they are now, it's safe to say that normal antediluvian humans were twice as big as they are today. The antediluvian giants were likely at least 20 feet tall.

Earths Evil Spirits

There is the possibility that all evil spirits on the Earth today are the spirits of the Nephilim. If the book of Enoch is correct, they are. We read about this in Enoch 15:

> 8. And now, the giants, who are produced from the spirits and flesh, shall be called evil spirits upon the earth, and on the earth shall be their dwelling. 9. Evil spirits have proceeded from their bodies; because they are born from men, and from the holy Watchers is their beginning and primal origin; they shall be evil spirits on earth, and evil spirits shall they be called. 10. As for the spirits of heaven, in heaven shall be their dwelling, but as for the spirits of the earth which were born upon the earth, on the earth shall be their dwelling. 11. And the spirits of the giants afflict, oppress, destroy, attack, do battle, and work destruction on the earth, and cause trouble: they take no food, but nevertheless hunger and thirst, and cause offences. And these spirits shall rise up against the

children of men and against the women, because they have proceeded from them.

If this is correct, then the Nephilim are responsible for much of the evil that we see on our Earth today. It is their mission to "rise up against the children of men and against the women." The Nephilim descended in the time of Jared (the fifth from Adam) and caused humans to sin more than they would have before.

Every one of us needs salvation from Jesus Christ. We can't be good enough to save ourselves, otherwise, why did Jesus have to come die? Ephesians 2:8-9 says that we can't do enough good to save ourselves, *"For by grace are ye saved through faith; and that not of yourselves: it is the gift of God: Not of works, lest any man should boast."*

If you're in need of a savior, turn your back on your sin (repent) and put your trust in Jesus alone. Acts 3:19 reads, *"Repent ye therefore, and be converted, that your sins may be blotted out, when the times of refreshing shall come from the presence of the Lord."*

John 3:16 reads, *"For God so loved the world, that he gave his only begotten Son, that whosoever believeth in him should not perish, but have everlasting life."*

Now is the time of repentance. Don't put it off.

Some speculate that Neanderthals were the Nephilim. Foutry-two percent of all humans today have Neanderthal DNA in their bodies. Every human with that DNA can trace their ancestry back to three women. Could those be Noah's three daughters-in-law? While I can't rule the theory out, the Bible is clear that the Nephilim were giants. Could Neanderthal skeletons be adolescent giants that never lived long enough to grow to their full stature? Could their full-grown giant skeletons have been hidden away or destroyed by the Smithsonian?

Could 42 percent of Earth's population have corrupted DNA from the watchers? Consider, our DNA is 98 percent similar to that of chimpanzees, however, they have less than 200 genetic disorders, but we have over 6,000.

In fact, there are numerous articles explaining traits and diseases that modern humans with Neanderthal DNA have. Those with the DNA are at a significantly increased risk of becoming addicted to nicotine. They are more prone to depression. Vanderbilt University Medical Center lists dozens of negative health conditions that these people have.

On the flip side, some humans may have the "god gene." In the early 2,000s geneticist Dean Hamer conducted research on religion and spirituality. Hamer speculated that faith was coded into us as humans and that those who have the god gene are more optimistic and recover from disease more quickly among other things.

In his experiments, Hamer determined that the more spiritual a person was, the more likely they were of having the gene VMAT2, a monoamine transporting gene (transporting chemicals in the brain). It was determined that this gene played a large role in a person's consciousness and how they perceive the world.

Most of us grow up being taught that demons are what we are fighting against. I theorize that there is more than one type of "bad" spiritual being we are facing. Demons are the spirits of the Nephilim whose sole purpose is to oppose mankind and make us miserable. Princes of the air are the more intelligent, more dangerous fallen angels who rule the nations. Satan, I believe, is one of these if not their leader. I look forward to expounding on this theory more in my fourth Marginal Mysteries series on angels.

Samson the Nephilim

There is evidence in scripture that Samson's may have been a Nephilim. Let's start by quickly summarizing the story of Samson in the book of Judges.

Samson's parents are visited by an angel who tells them that Samson would be a Nazarite all his life. This meant Samson couldn't eat certain foods, touch a dead body, or cut his hair. In exchange for these sacrifices, God endows Samson with exceptional strength.

Samson falls in love with a Philistine woman. The Philistines are the enemies of Samson and the people of God. Samson kills a lion with his bare hands on one of his trips to meet with this woman. On the way back home, he scoops honey out of the lion's carcass thus violating his Nazarite oath to not touch a dead body.

After the Philistines burned Samson's wife, he singlehandedly went to war against the Philistines and killed many of them in a "great slaughter." Judges 15:8 reads, *"And he smote them hip and thigh with a great slaughter: and he went down and dwelt in the top of the rock Etam."*

He turned himself over to the Philistines. When Samson got to the Philistine camp, he broke his ropes and slew 1,000 Philistines with a donkey's jawbone. Judges 15:15-16 reads, *"And he found a new jawbone of an ass, and put forth his hand, and took it, and slew a thousand men therewith. And Samson said, With the jawbone of an ass, heaps upon heaps, with the jaw of an ass have I slain a thousand men."*

Samson became a judge of Israel for twenty years. He fell in love with a woman from Gaza. Samson was surrounded by the Gazites who wished to kill him, but he escaped by ripping the city gates off their hinges. We read in Judges 16:3, *"And Samson lay till midnight, and arose at midnight, and took the doors of the gate of the city, and the two posts, and went away with them, bar and all, and put them upon his shoulders, and carried them up to the top of an hill that is before Hebron."*

Does this passage also imply that Samson was a giant? At the very least, Samson had to be a big man to fit the city gate on his shoulders.

Again, Samson fell in love with a woman. She was a Philistine named Delilah. The Philistine leaders promised to pay Delilah if she could discover the source of Samson's strength. She discovers that Samson's uncut hair is the source of his strength, so she cuts it off while he's asleep. The Philistines capture Samson who has now lost his strength and gouge out his eyes according to Judges 16:21, *"But the Philistines took him, and put out his eyes, and brought him down to Gaza, and bound him with fetters of brass; and he did grind in the prison house."*

Notice that Samson "did grind" in the prison house. The Hebrew word for grind is *"ṭāḥan"* which means to grind meal or to be a concubine. The word *ṭāḥan* is used in Job 31:10 to describe a concubine, *"Then let my wife grind unto another, and let others bow down upon her."*

If in fact Samson were a concubine in the prison of the Philistines, it would insinuate that a lot of women wanted to have a child of super-human strength or even size by him.

While in captivity, Samson's hair grew back. When the Philistines were having a feast, they brought up Samson to make fun of him. Samson asked the boy leading him to bring him to the main pillars of the building so that he could rest against them. Samson asked the Lord for strength to push the pillars down. The Lord granted Samson's request and Samson killed himself and everyone in the gathering – more than in his entire life, according to Judges 16:29-30, *"And Samson took hold of the two middle pillars upon which the house stood, and on which it was borne up, of the one with his right hand, and* of the other with his left. And Samson said, Let me die with the Philistines. And he bowed himself with all his might; and the house fell upon the lords, and upon all the people that were therein. So the dead which he slew at his death were more than they which he slew in his life."

The fact that Samson was able to put a hand on each pillar also indicates that he was a man of great size if not a giant. From his story, we certainly see that Samson was a mighty man of renown – the Genesis 6:4 description of the Nephilim, "There were giants in the earth in those days; and also after that, when the sons of God came in unto the daughters of men, and they bare children to them, the same became mighty men which were of old, men of renown."

Did an angel come into Samson's mother to conceive him? Let's look at Judges 13. Samson's mother, who is unnamed, is not able to have children. Unlike Sara, Rebecca, or Hannah in scripture, Samson's mother never prays for a child. In Judges 13:2-3, an angel appears to her, "And there was a certain man of Zorah, of the family of the Danites, whose name was Manoah; and his wife was barren, and bare not. And the angel of the Lord appeared unto the woman, and said unto her, Behold now, thou art barren, and bearest not: but thou shalt conceive, and bear a son."

The woman tells her husband, Manoah about the angel. Manoah wants to talk with the angel to see how the child should be properly raised. God hears Manoah and sends the angel back in Judges 13:9-10, "And God hearkened to the voice of Manoah; and the angel of God came again unto the woman as she sat in the field: but Manoah her husband was not with her. And the woman made haste, and ran, and shewed her husband, and said unto him, Behold, the man hath appeared unto me, that came unto me the other day."

The Hebrew word for "came" in "came unto me" is the word "**bô**'". It means to come in, go in, or go. This is the same word used in Genesis 6:4 when the sons of God came into the daughters of men,

"There were giants in the earth in those days; and also after that, when the sons of God came in unto the daughters of men, and they bare children to them, the same became mighty men which were of old, men of renown."

It's also used in Deuteronomy 22:13 for the phrase "and go in." If any man take a wife, and go in unto her, and hate her,

Also here in Judges 16:1, in the very story of Samson *bô'* is used for, again, "and went in": Then went Samson to Gaza, and saw there an harlot, and went in unto her.

Also, notice how the angel appeared to her in the field. There are places in scripture where fields are used for intercourse. Song of Solomon 7:12 reads, *"Let us get up early to the vineyards; let us see if the vine flourish, whether the tender grape appear, and the pomegranates bud forth: there will I give thee my loves."*

Deuteronomy 22:25,27 reads: *"But if a man find a betrothed damsel in the field, and the man force her, and lie with her: then the man only that lay with her shall die... For he found her in the field, and the betrothed damsel cried, and there was none to save her."*

We have more evidence that Samson's father may have been an angel. The name, Samson, implies that he is connected to the sun-god. The Hebrew word **"shemesh"** means sun. Samson likely means of the sun or like the sun or sun child. As we read in scripture, angels are associated with stars and even called stars. Revelation 1:20 says that the angels are stars in a vision, *"The mystery of the seven stars which thou sawest in my right hand, and the seven golden candlesticks. The seven stars are the angels of the seven churches: and the seven candlesticks which thou sawest are the seven churches."*

Manoah was not smart nor a leader. While it's possible that he was Samson's father, it would be a weird relationship between the weak father and his super-strength son.

The story of the Roman demigod Hercules (Heracles in Greek) appears to have been borrowed from the Hebrew story of Samson. Hercules was famous for his strength and vast adventures. He was the son of the Roman god Jupiter and mortal woman Alcmena.

The Egyptian version of Hercules is named Som, which is closer to Samson. In the Egyptian version of his death, Som being brought from prison to be sacrificed, gathers his strength, and massacres many thousands of spectators before his death.

The idea that Samson may have been conceived of an angel is an interesting study. As stated earlier, the word, *bô'* can also mean to go, so it is possible that my speculation is wrong. The speculation and some evidence are there nonetheless.

Nephilim in Secular History

The Nephilim are found in many ancient stories including Mesopotamian, Sumerian, Acadian and Babylonian stories. In these they go by the Anunnaki, Igigi and Sebitti. Giants appear in legends across the globe. In the United States, giant skeletons have been found across the country. Most native American tribes have stories about giants that they struggled with.

Why did Native Americans greet each other by showing their hands? It was to see if you had giant's blood running through your veins.

Polydactyly is a condition in which someone is born with one or more extra fingers. It was a common condition among the giants. We read in 1 Chronicles 20:6, *"And yet again there was war at Gath, where was a man of great stature, whose fingers and toes were four and twenty, six on each hand, and six on each foot and he also was the son of the giant."*

The Old World Roots of the Cherokee is a book that traces the origins of the Cherokee people to the third century B.C. It follows their migrations through the Americas to their homeland in the Appalachian Mountains. The book claims that the Cherokee Indians are descended from Jewish and Eastern Mediterranean people who spoke Greek before adopting the Iroquoian language of their allies, the Haudenosaunee, in the Ohio Valley. In the book, we learn why the Cherokee showed their hands as a greeting. Old World Roots of the Cherokee reads:

> What kind of Indians lived in the territory the Choctaw and Chickasaw carved out for their new home? According to their traditions, as confirmed by excavations of bones in Tennessee, it was a "race of white giants": "Hau" is a native American greeting used by many tribes such as the; Sue, the Tetons, the Dakota, and the Omaha and it was to see how many fingers you had on your hand.

Cory Daniel is a native American historian that writes:

> The showing or shaking of hands came about to show that one was not of tainted bloodline. Five fingers proved that one did not possess the traits of blood-lust and violence which always accompanied the descendants of six-fingered giants, or "men of renowned"

Before we leave the Appalachian Mountains, let's also discuss the Melungeons. They were discovered in the Appalachian Mountains in 1654 by English explorers and were said to be neither of Indian nor Negro descent but had European features.

Like the Cherokee, the Melungeons are said to have been of Mediterranean descent and settled in the Appalachian wilderness. These people intermarried with six native races including the Cherokee.

Hold the book with one hand and with the other, feel the back of your head where it meets the neck. If you have a half golf-ball size bump, then you may be of Melungeon descent. This is also a trait of people from the Anatolian region of Turkey.

Descendants of the Melungeons, along with those of the Turkish Altl parmak family can have six fingers or toes. The name Altl parmak literally means "six fingered ones."

A fossilized 12-foot 2-inch tall giant was found in Ireland in 1895. He had six toes on his right foot. The remains disappeared on a railway trip between being exhibited in England.

Also, in Ireland lived Charles Byrne nicknamed "The Irish Giant". His 8-foot 4-inch tall skeleton was on display at the Hunterian Museum in London from 1799 until its removal in 2023.

Left: Fossilized giant found in a bog in 1895
Right: The skeleton of Charles Byrne Credit: Emoke Denes, Wikipedia

A six fingered giant skeleton found on Catalina Island in the early 1900s

Giant skeletons have been found all over the United States. Legends of giants permeate Native American history. Over a dozen petroglyphs in the southwestern United States depict hands and feet with six digits on each.

National Geographic reported in 2016 that a Pueblo culture living in the Chaco Canyon of New Mexico gave exalted status to those among them that had six toes on each foot. Out of 96 skeletal remains found in the canyon, three individuals had six digits on their feet.

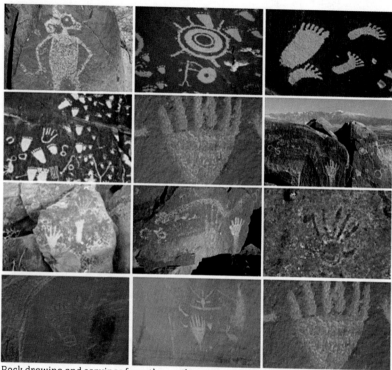

Rock drawing and carvings from the southwest United States

In the many Buffalo Bill stories we read about life in the old West, one excerpt from The Mounted Sharps of the Overland reads:

> The one they spoke of as the Giant Miner he had heard of-had, in fact, seen several times. He knew him as a man of giant form, who dwelt alone in the mountains, going to the camp only when he needed provisions, and constantly on the search for gold. The Indians were afraid of him, the outlaws left him alone, and yet he was considered harmless...

Buffalo Bill shares a fascinating and relevant Pawnee story in his autobiography:

> While we were in the sandhills, scouting the Niobrara country, the Pawnee Indians brought into camp some very large bones, one of which the surgeon of the expedition pronounced to be the thigh bone of a human being. The Indians said the bones were those of a race of people who long ago had lived in that country. They said these people were three times the size of a man of the present day, that they were so swift and strong that they could run by the side of a buffalo, and, taking the animal in one arm, could tear off a leg and eat it as they ran. These giants, said the Indians, denied the existence of a Great Spirit. When they heard the thunder or saw the lightning, they laughed and declared that they were greater than either. This so displeased the Great Spirit that he caused a deluge. The water rose higher and higher till it drove these proud giants from the low grounds to the hills and thence to the mountains. At last even the mountaintops were submerged and the mammoth men were drowned. After the flood subsided, the Great Spirit came to the conclusion that he had made men too large and powerful. He therefore corrected his mistake by creating a race of the size and strength of the men of the present day. This is the reason, the Indians told us, that the man of modern times is small and not like the giants of old. The story has been handed down among the Pawnees for generations, but what is its origin no man can say.

Kap-Dwa is a Patagonian giant discovered off the coast of South America in 1673. He is a two-headed 12-foot-tall giant. Not only do we have pictures, but his mummified corpse is on display in Baltimore, Md., at an oddities shop called Bob's Side Show.

Kap-Dwa, the two headed, 12-foot-tall giant.

Eight-foot-tall red headed giants were found in Lovelock Cave, Nevada in 1911. Four of their giant skulls were displayed in a museum near the cave until around 2010.

Left: Lovelock Cave | Right: Winnemucca with real red giants' hair in her necklace

A woman named Sarah Winnemucca, descendant of Chief Winnemucca of the Piute tribe, wrote a book in 1882 recounting a battle between her people and a race of giants known as the sai-duka. The sai-duka were red-haired cannibal giants that lived in the mountains.

The Piute people grew tired of being cannibalized by these giants so went to war. The last of the giants held up in Lovelock Cave. The Piute's piled brush and wood at the cave entrance and killed the last of the giants. You can visit Lovelock Cave today and see the scorch marks at the cave entrance.

Ancient Monuments

Many of our ancient structures couldn't be built today even with our advanced technology. Many of the stones in these structures weigh too much for our biggest mobile cranes to lift. Structures around the globe have rocks with impossibly precise angles cut into them by an unknown technology. Who built the great structures of our ancient world and what technology did they use to do so?

The Great Pyramid of Giza has amazing mathematics involved in its proportions and measurements. Without getting into a dozen other mathematical anomalies, I'll mention one of the more amazing ones. You can draw a circle inside of the pyramid's triangle and the circle will have a diameter of 365.242 feet. What else is exactly 365.242? That's the exact number of days in a year to the thousandths.

Stone of the Pregnant Woman

The Stone of the Pregnant Woman is the largest known cut stone in the world weighing over 2,000 tons (estimated). Three other smaller stones under The Grand Terrace of the Temple of the Sun at Ba'albek were quarried from the same location and transported over a half mile uphill to the temple. They weigh between 750 and 1,000 tons each.

Machu Picchu is an Incan ruin high up in the Andes with steep drops on three sides. Some of the temple stonework is so well-placed together that there are no gaps in any of the joints. These stones are placed together without mortar. One Incan stone at a different location known as the twelve-angled stone, has 12 different angles carved into its border.

Gobekli Tepe is believed to be the site of the world's oldest known megaliths. Pillars at the site are decorated with details of our ancient Earth including clothing and wild animals.

Gobekli Tepe is found within the Tas Tepeler region of Anatolia, Turkey, along with 11 other ancient sites. At these sites we find multiple engravings of humans with six fingers. These sites also claim the trophy for the oldest phallus painting (lots of them) in the world secularly dated to 27,000 years ago.

You'll notice at many of these ancient sites that cruder building techniques are used on top of better techniques. It's almost as if more advanced beings built many of these monuments long ago and newer less advanced civilizations came after and claimed the monuments as their own.

Did the Nephilim with advanced knowledge from the Watchers build our ancient megaliths? Whatever the answer, the ancients had far more advanced knowledge than most believe.

Modern Nephilim

Could giants still exist today? There are accounts of them existing. Is the book of Enoch correct when it says that the generation of the Great Tribulation will understand what's contained in the book?

Robert Wadlow was an American from Illinois who was the tallest person in modern recorded history measuring 8 feet 11 inches tall. His stature was due to a pituitary tumor.

Sultan Kösen is currently the tallest living man at 8 feet 3 inches. He was born in Turkey in 1982. His stature is due to a pituitary tumor.

Chang Woo Gow was a 7 feet 9 inches tall man born in Canton Province China in 1841. Ondor Gondor was a Mongolian who also measured 7 feet 9 inches tall.

Left: Robert Wadlow | Middle: Chang Woo Gow | Right: Ondor Gongor

Modern gigantism is usually a result of ailments that cause excessive production of HGH, or human growth hormone. In some cases, it's because of tumors in the brain. In almost every instance today, these giants have severe health problems and often die at young ages. The giants in scripture are warriors and mighty men of valor, therefore likely didn't suffer from ailments that caused their size. The gigantism in scripture is hereditary.

There are multiple reports of giants being seen in Afghanistan by U.S. troops. Keep in mind that in that region, near Mongolia, we already have plenty of pictures of giants – some of which are in this book.

Afghanistan is most definitely a place out of time. On my deployment to Afghanistan, I was able to visit one of Alexander

Micah at Alexander the Great's castle in Herat Afghanistan

Once during my deployment to Afghanistan, I flew from Kabul to Herat. On my low flying aircraft's flight, I remember seeing nothing but rugged barren mountains as far as I could see. I took a 45-minute nap and when I woke up, I was still looking at barren mountains as far as I could see. Though I didn't see any giants myself, it was obvious that mankind likely had not set foot on those mountains – ever.

One unknown U.S. soldier recounts his experience of seeing a 12-foot giant on one mountain near Kunar, Afghanistan:

> My lieutenant gave me a new thermal imaging system called the Recon3 that none of us were familiar with and told me to figure out what I can and pass along that information to the other team leaders. I started messing with the Recon3 to see its capabilities and was surprised at the clarity of the zoom on it. I spent most of my time messing with the different functionalities and watching the village.
>
> I started to look across the valley to what I could see and that led me to look along the spur we were set in on and saw a very large heat signature at the top of one of the false peaks. I did everything I could to get as clear of an image as I could, suspecting that it was a group of Taliban huddled together around a light as they tend to do in the mountains. All of the sudden, the heat signature

stood up as one being. The trees in that area grew up to about 10 to 12 feet tall, and this thing was at least as tall, if not taller than the trees that surrounded it. It started taking steps parallel to my position and was covering ground quickly with ease. Its stride was slow and relaxed, yet it moved with incredible speed. That led me to believe that this creature was gigantic.

...I didn't tell many people about it, while I was in and even when I got out. I kept it to myself, thinking there was no way I saw what I saw. But then in 2010, I listened to a story on Coast to Coast... specifically the story about the Giant of Kandahar. That made all the memories of my time in service come flooding back and made me consider other things I saw during that deployment.

For instance, the creature was described as having fire-orange hair and it reminded me of a tradition of locals in the area of my sighting. They would dye their hair a bright orange color and even would dye their goats the same color. They never gave any explanation why. It seemed like it was every once in a while they would do this, and then all of a sudden those orange-dyed goats would be gone, and the locals' hair would also no longer be dyed orange. I assumed maybe it was a cultural thing I didn't understand, but now it makes me wonder if that was some kind of gesture to the creature/Nephilim or if the goats were sacrificed to it.

I am a Christian, and the Bible briefly discussed the "Men of Renown," a.k.a. the Nephilim. I think that's what I saw, a member of an ancient race of giants that descended from fallen angels. Or it could be something like Sasquatch. I'm not sure.

Before we talk about the giant of Kandahar, I'll back up some of what this soldier is saying. Indeed, in 2006, my sniper platoon in Fallujah, Iraq was given one thermal scope. We had five teams, so we didn't always get to bring it with us on missions. When it was my team's (Redneck 4) turn, I always asked to take night watch because the technology was amazing. I remember watching field mice running around 200 yards away.

Chief among these stories is that of the giant of Kandahar. A special forces team came upon a cave high up on a mountain. The ground around the cave was littered with bones and a horrible stench emanated from the cave. A 13-foot-tall red-headed man charged out of the cave and speared Dan, an unfortunate soldier.

The soldiers dispatched the giant and radioed it in. Upon further inspection of the corpse, the soldiers said that he had six fingers on each hand and double rows of teeth. A chinook helicopter flew in, picked up the giant in a net, and flew off with it.

U.S. Forces extract in Afghanistan., ResoluteSupportMedia on Flickr

The soldiers were required to sign non-disclosure agreements. That's easy for me to believe as I was required to sign a non-disclosure agreement for one of my missions in Fallujah, Iraq.

Steve Quayle is an author and radio host who hosted Survive2thrive and Coast to Coast. He interviewed a pilot who flew a dead giant out of Afghanistan in 2005. The giant of Kandahar was slain in 2002, meaning that multiple giants have been killed by U.S. forces.

L.A. Marzulli interviewed a soldier who told him of the rumors around the base in Kandahar about the giant. The soldier also participated in cave training and was instructed to "aim high." Is the story of giants in the Afghanistan mountains true? It's plausible.

CHAPTER 8

THE POSTDILUVIAN EARTH

After the flood, God disperses the nations at the Tower of Babel. Most of us have been taught that the people started to build the tower at Babel to reach God. Genesis 11, tells us that they built the tower to make a name for themselves so as not to be scattered abroad:

> ¹ And the whole earth was of one language, and of one speech. ² And it came to pass, as they journeyed from the east, that they found a plain in the land of Shinar; and they dwelt there. ³ And they said one to another, Go to, let us make brick, and burn them thoroughly. And they had brick for stone, and slime had they for morter. ⁴ And they said, Go to, let us build us a city and a tower, whose top may reach unto heaven; and let us make us a name, lest we be scattered abroad upon the face of the whole earth.

The people that journeyed west attempted to build a tower to inspire mankind to congregate into a city. Notice that this is in direct disobedience to God's command to mankind to go throughout all the earth and multiply in Genesis 9:7, *"And you, be ye fruitful, and multiply; bring forth abundantly in the earth, and multiply therein."* ...continued in verse 19, *"These are the three sons of Noah: and of them was the whole earth overspread."*

Because of the people's disobedience, God confounds their language in order to separate them in Genesis 11:

> ⁵ And the Lord came down to see the city and the tower, which the children of men builded. ⁶ And the Lord said, Behold, the people is one, and they have all one language; and this they begin to do: and now nothing will be restrained from them, which they have imagined to do. ⁷ Go to, let us go down, and there confound their

language, that they may not understand one another's speech. ⁸So the Lord scattered them abroad from thence upon the face of all the earth: and they left off to build the city. ⁹Therefore is the name of it called Babel; because the Lord did there confound the language of all the earth: and from thence did the Lord scatter them abroad upon the face of all the earth.

The Princes of the Air

After God scatters mankind at Babel, He divides the world into at least 70 nations. This is the point where God chooses Israel as his nation. What is fascinating is I believe that God gives the other 70 nations to angels to rule over. The number of the children of Israel is 70 as seen in Genesis 46:27, *"And the sons of Joseph, which were born him in Egypt, were two souls: all the souls of the house of Jacob, which came into Egypt, were threescore and ten."*

We read in Deuteronomy 32 about the dispersion of the nations:

⁷Remember the days of old, consider the years of many generations: ask thy father, and he will shew thee; thy elders, and they will tell thee. ⁸When the Most High divided to the nations their inheritance, when he separated the sons of Adam, he set the bounds of the people according to the number of the children of Israel. ⁹For the Lord's portion is his people; Jacob is the lot of his inheritance.

Here in the Masoretic text, God sets up 70 nations. In the Greek Septuagint, verse 8 reads that God gives the nations to the "angelos theos" which is translated the angels of God. In the Dead Sea Scrolls, verse 8 reads that God gives the nations to the bene ha Elohim which are the sons of God, or the Watchers!

Does God set up the angels as the rulers of the 70 nations? I believe that He does. I also believe that He either gave some nations to fallen angels or these angels fell after this division of the nations. It's backed up in scripture many times. As we already touched on in the chapter on the Watchers, we read about the angelic rulers of the nations, or princes, in Daniel 10, "*13 But the prince of the kingdom of Persia withstood me one and twenty days: but, lo, Michael, one of the chief princes, came to help me; and I remained there with the kings of Persia. ...20 Then said he, Knowest thou wherefore I come unto thee? and now will I return to fight with the prince of Persia: and when I am gone forth, lo, the prince of Grecia shall come.*"

We see in verse 20 that there is a prince of Greece. Michael is the prince of Israel according to Daniel 12:1, "*And at that time shall Michael stand up, the great prince which standeth for the children of thy people...*"

Psalm 82:6-7 tells us that the princes fell, "*I have said, Ye are gods; and all of you are children of the most High. But ye shall die like men, and fall like one of the princes.*"

Psalm 96:4-5 calls the gods of the nations, idols, "*For the Lord is great, and greatly to be praised: he is to be feared above all gods. For all the gods of the nations are idols: but the Lord made the heavens.*"

It's not just the Bible that says that God divided up the world into different nations. As we studied in my last book, *Ancient Cities and the gods Who Built Them*, the dividing up of the nations amongst the gods is the first thing mentioned in the story of Atlantis. Plato writes in the Critias:

> In the days of old the gods had the whole earth distributed among them by allotment. There was no quarrelling; for you cannot rightly suppose that the gods did not know what was proper for each of them to have, or, knowing this, that they would seek to procure for themselves by contention that which more properly belonged to others. They all of them by just apportionment obtained what they wanted...

In the Sumerian *Eridu Genesis*, each of the world's cities are given to a god to oversee. Fragments of the tablet are missing so the translation is broken up:

> After the...of kingship had descended from heaven, after the exalted crown and throne of kingship had descended from heaven, the divine rites and the exalted powers were perfected, the bricks of the cities were laid in holy places, their names were announced and the...were distributed. The first of the cities, Eridu, was given to Nudimmud the leader. The second, Bad-tibira, was given to the Mistress. The third, Larag, was given to Pabilsag. The fourth, Zimbir, was given to hero Utu. The fifth, Suruppag, was given to Sud. And after the names of these cities had been announced and the...had been distributed, the river...was watered. [Here there are about 34 lines missing]

In the Egyptian *Book of the Holy Cow*, Ra leaves the lesser gods to rule over the Earth. We read in the *Book of the Holy Cow*:

> The Majesty of this god said, "Stay far away from them! humanity Lift me up! Look at me!" and so Nut became the sky Then the Majesty of this god was visible within her. She said, "If only you would provide me with a multitude to help me!" and so the Milky Way came into being

Again, as it's relevant here, the idea of the gods meeting to discuss the affairs of the Earth and mankind is recorded in scripture. Hebrew words associated with the heavenly council of the gods include council, assembly, congregation, host, the holy ones and sons of God. Following is a list of some of the mentions of this council of the gods in scripture.

> **Psalm 82:1,** God standeth in the congregation of the mighty; he judgeth among the gods.

> **Job 2:1,** Again there was a day when the sons of God came to present themselves before the Lord, and Satan came also among them to present himself before the Lord.

Psalm 89:5, And the heavens shall praise thy wonders, O Lord: thy faithfulness also in the congregation of the saints.

1 Kings 22:19, ...I saw the Lord sitting on his throne, and all the host of heaven standing by him on his right hand and on his left.

Daniel 4:17, This matter is by the decree of the watchers, and the demand by the word of the holy ones: to the intent that the living may know that the most High ruleth in the kingdom of men, and giveth it to whomsoever he will, and setteth up over it the basest of men.

God says that you should not have any other gods before Him in Exodus 20:2-4, *"I am the Lord thy God, which have brought thee out of the land of Egypt, out of the house of bondage. Thou shalt have no other gods before me. Thou shalt not make unto thee any graven image, or any likeness of any thing that is in heaven above, or that is in the earth beneath, or that is in the water under the earth."*

I believe that when God says to the children of Israel that they should have no other gods before Him, that he truly meant physical beings and/or the images of those beings after they had died or were no longer manifest. There are at least 40 heathen gods mention in our Old Testament including: Adrammelech, Amun, Anat, Anammelech, Asherah, Ashima, Astaroth, Astarte, Baal, Baal Berith, Baal Peor, Baal-zephon, Beelzebub, Bel, Belial, Dagon, Dumuzid, Gad, Hauron, Inanna, Kamos, Marduk, Melqart, Milcom, Moloch, Mot, Nabu, Nehushtan, Nergal, Nibhaz, Ninurta, Nisroch, Remphan, Resheph, Rimmon, Shahar, Shapash, Succoth-benoth and Yarikh.

In 2 Kings 10, the Israelite King Jehu tricked all the servants of Baal to come to a great sacrifice where he had his soldiers massacre them. Jehu then made the house of Baal into a public bathroom. 2 Kings 10:27 reads, *"And they brake down the image of Baal, and brake down the house of Baal, and made it a draught house unto this day."*

In 1 Samuel 5, the Philistines captured the Ark of the Covenant which they placed in the temple of Dagon. Two days in a row, the statue of Dagon fell face-down before the ark and broke. The Dagon worshipers, worrying, moved the ark from town to town but the people developed tumors wherever the ark went. The Philistines, tired of the curse, returned the ark to Israel. 1 Samuel 5:11 reads, *"So they sent and gathered together all the lords of the Philistines, and said, Send away the ark of the God of Israel, and let it go again to his own place, that it slay us not, and our people: for there was a deadly destruction throughout all the city; the hand of God was very heavy there."*

The story of Nisroch is a fascinating one. Jewish legends tell of him actually being a plank of wood from Noah's ark. Sennacherib worshiped the plank of wood because he realized that it must have been the god that saved Noah from the flood. It's quite possible that Sennacherib had been to Mount Ararat considering his assassins (his sons) escaped to Ararat. He named the wood god Nisroch. We read of Sennacherib's assassination in Isaiah 37:38, *"And it came to pass, as he was worshipping in the house of Nisroch his god, that Adrammelech and Sharezer his sons smote him with the sword; and they escaped into the land of Armenia: and Esarhaddon his son reigned in his stead."*

The Hebrew for "Armenia" is **'ărārāṭ**. The scribes of the Masoretic likely changed Ararat to Armenia because at the time of translation, it was known as the Land of Armenia and not Ararat.

Are Cities Bad?

As we read earlier in this chapter, after the flood, mankind was told to spread out and fill the Earth, but the people at Babel built a city so that they could be gathered together. Cities can easily be seen as monuments to man. They exist for man and represent his achievements. In the instance of Babel, the city was a bad thing. Lot chose to live in Sodom where great wickedness took place. Enoch 69:21 mentions filling the Earth with cities as a bad thing, "Prepare slaughter for his children Because of the iniquity of their fathers, Lest they rise up and possess the land, And fill the face of the world with cities."

In Isaiah 14:21-22, God is sets out to destroy Babylon, *"Prepare slaughter for his children for the iniquity of their fathers; that they do not rise, nor possess the land, nor fill the face of the world with cities. For I will rise up against them, saith the Lord of hosts, and cut off from Babylon the name, and remnant, and son, and nephew, saith the Lord."*

One great visualization from the terrible 2014 movie Noah was the image that the world had become one big city that had consumed the land. Here in the United States, our cities are where the vast majority of violent crimes take place. Political affiliation is polarized between the rural and urban.

Since my time serving as a Tennessee State Representative, I've pondered some questions. Do cities attract liberals or do cities create liberals? Do rural areas attract conservatives or do rural areas create conservatives? It's probably a combination of attraction and creation but I still don't have a good answer for that.

The Aesop's fable, The Town Mouse and the Country Mouse sums the idea up. The town mouse visits the country mouse and thinks the lifestyle is rather bland. The country mouse visits the town mouse and almost dies three times while enjoying the delicacies of city life. In the end, the country mouse says, "You may have luxuries and dainties that I have not, she said as she hurried away, but I prefer my plain food and simple life in the country with the peace and security that go with it."

The first city mentioned in the Bible was built by Cain. Cain, of course, is the first murderer in history. Genesis 4:16-17 reads: *"And Cain went out from the presence of the Lord, and dwelt in the land of Nod, on the east of Eden. And Cain knew his wife; and she conceived, and bare Enoch: and he builded a city, and called the name of the city, after the name of his son, Enoch."*

This is not the same Enoch that supposedly wrote the book of Enoch– that Enoch was from the bloodline of Seth.

After the flood, Noah's son Ham is cursed for making fun of his fathers' nakedness. Canaan is the son of Ham in Genesis 9:

> [24] And Noah awoke from his wine, and knew what his younger son had done unto him. [25] And he said, Cursed be Canaan; a servant of servants shall he be unto his brethren. [26] And he said, Blessed be the Lord God of Shem; and Canaan shall be his servant. [27] God shall enlarge Japheth, and he shall dwell in the tents of Shem; and Canaan shall be his servant.

Notice that Japheth and Shem dwell in tents. Ham is the only son whose bloodline is mentioned as building cities. We read in Genesis 10:10-12, *"And the beginning of his kingdom was Babel, and Erech, and Accad, and Calneh, in the land of Shinar. Out of that land went forth Asshur, and builded Nineveh, and the city Rehoboth, and Calah, And Resen between Nineveh and Calah: the same is a great city."*

Jonah was told to go preach against the city of Nineveh because of its wickedness. The entire world needs the gospel. How does that fit in with these questions? Everyone is in need of a Savior no matter where they live.

The Return of the Firmament

There is much debate among Christians as to the end-times timeline. While it's not my goal to join in that debate in this book, I will speculate as to some of the events in that timeline.

I believe that the new earth prophesied about in the book of Isaiah will take place after the seven-year tribulation, the second coming, and the judgment of the nations. During the tribulation, the Earth will undergo such turmoil that it will need to be recreated for the millennium. Revelation 16:3-4 tells us about some of this turmoil, *"And the second angel poured out his vial upon the sea; and it became as the blood of a dead man: and every living soul died in the sea. And the third angel poured out his vial upon the rivers and fountains of waters; and they became blood."*

All of the water on the Earth becomes blood and every living creature in the waters die. After this, every island and mountain will be changed. We read in Revelation 16:20, *"And every island fled away, and the mountains were not found."*

If indeed Isaiah 24:19-20 is speaking of the Great Tribulation instead of Sennacherib's Assyrian army or Nebuchadnezzar's army, then it foretells great destruction, *"The earth is utterly broken down, the earth is clean dissolved, the earth is moved exceedingly. The earth shall reel to and fro like a drunkard, and shall be removed like a cottage; and the transgression thereof shall be heavy upon it; and it shall fall, and not rise again."*

Isaiah 34:4 talks of the heavens being rolled back as a scroll, *"And all the host of heaven shall be dissolved, and the heavens shall be rolled together as a scroll: and all their host shall fall down, as the leaf falleth off from the vine, and as a falling fig from the fig tree."*

In Revelation 19:11, John sees this event, *"And I saw heaven opened, and behold a white horse; and he that sat upon him was called Faithful and True, and in righteousness he doth judge and make war."*

Peter tells us that the heavens will pass away the day the Lord returns in 2 Peter 3:10-11, *"But the day of the Lord will come as a thief in the night; in the which the heavens shall pass away* with a great noise, and the elements shall melt with fervent heat, the earth also and the works that are therein shall be burned up. Seeing then that all these things shall be dissolved, what manner of persons ought ye to be in all holy conversation and godliness"

Psalm 97:4-5 describes the day too, "His lightnings enlightened the world: the earth saw, and trembled. The hills melted like wax at the presence of the Lord, at the presence of the Lord of the whole earth."

The heavens pass away and the Earth is destroyed. This should make a new Heaven and a new earth necessary for the Millennium. I believe Isaiah 65 is speaking of the Millennium:

> [7]For, behold, I create new heavens and a new earth: and the former shall not be remembered, nor come into mind. ...[21]And they shall build houses, and inhabit them; and they shall plant vineyards, and eat the fruit of them. ...[23]They shall not labour in vain, nor bring forth for trouble; for they are the seed of the blessed of the Lord, and their offspring with them. ...[25]The wolf and the lamb shall feed together, and the lion shall eat straw like the bullock: and dust shall be the serpent's meat. They shall not hurt nor destroy in all my holy mountain, saith the Lord.

Isaiah 66:23 says that there will be a new moon, "And it shall come to pass, that from one new moon to another, and from one sabbath to another, shall all flesh come to worship before me, saith the Lord"

We skipped verses 20 and 22 in the chapter 65 passage because I want you to see something fascinating. Isaiah 65 reads:

> [20]There shall be no more thence an infant of days, nor an old man that hath not filled his days: for the child shall die an hundred years old; but the

sinner being an hundred years old shall be accursed. 22 They shall not build, and another inhabit; they shall not plant, and another eat: for as the days of a tree are the days of my people, and mine elect shall long enjoy the work of their hands.

Did you notice that when someone dies at 100 years old, they will still be a child? Did you notice that men's days are as the days of a tree? That's right! Men's lives will likely be as long as they were before the flood of Noah. I believe we can all look forward to the return of the firmament!

The Coming Nephilim?

The first thing that the book of Enoch says is that humans will not understand what is contained in the book until the last generation, the generation of the Tribulation. We read in Enoch 1:1-2, *"The words of the blessing of Enoch, wherewith he blessed the elect and righteous, who will be living in the day of tribulation, when all the wicked and godless are to be removed. And he took up his parable and said--Enoch a righteous man, whose eyes were opened by God, saw the vision of the Holy One in the heavens, which the angels showed me, and from them I heard everything, and from them I understood as I saw, but not for this generation, but for a remote one which is for to come."*

Does this mean that God will blur our minds as we read the book of Enoch or does it mean that the last generation, the generation of the Great Tribulation will see the things written about in the book (e.g., giants, chimeras), on the Earth at the end of times?

What is the beast in in the books of Daniel and Revelation? How is the image to the beast brought to life by someone other than God? Why is it an image? Does it not have a soul? Is this beast a chimera? We read in Revelation 13:14-15, *"And deceiveth them that dwell on the earth by the means of those miracles which* he had power to do in the sight of the beast; saying to them that dwell on the earth, that they should make an image to the beast, which had the wound by a sword, and did live. And he had power to give life unto the image of the beast, that the image of the beast should both speak, and cause that as many as would not worship the image of the beast should be killed."

While I don't know if the Nephilim will return, the theory doesn't go against what we've read in scripture. I'll leave it with the words of Jesus himself in Matthew 24:37, *"But as the days of Noah were, so shall also the coming of the Son of man be."*

CHAPTER 9

CONCLUSION

Elohim created a fantastical planet over which He placed mankind as caretakers. Angels interfered with the creation and their offspring utterly corrupted it. Elohim regretted that He had created man so He sent a great flood to wipe out the corruption.

Because of Adam's sin, we are all condemned to die. From the beginning, mankind has needed a savior. The Creator in His mercy sent His son Jesus as a sacrifice to pay our debt of death. While our mortal bodies will pass from the creation, our spirits will live for eternity. Because of Jesus' sacrifice, we can live with Him in Glory instead of apart from Him in Hell. Repent of your sin and trust Jesus for salvation.

If you earnestly seek the Creator, He will reveal Himself to you. Though He is not as physically present in the creation as He once was, He has left for us His Word in the Bible. Read it, find truth, wisdom, and salvation.

As buildings are a testament to a builder, creation is a testament to the Creator. Evolution is a hindrance to intelligent science. The theory of evolution is based on the idea that everything is getting better, that we are getting smarter and bigger. What the record shows is quite the opposite.

While modern medicine and nutritional understanding has given us longer lifespans, genetic mutations and diseases continue to manifest. Our bloodline was tainted with that of the Watchers. The evil spirits of the Nephilim bring disease and suffering. Romans 8:21-22 puts the fall of Genesis 3 into perspective, *"Because the creature itself also shall be delivered from the bondage of corruption into the glorious liberty of the children of God. For we know that the whole creation groaneth and travaileth in pain together until now."*

We are a species and culture on the brink of ushering in the Great Tribulation. The imagination of our hearts is only evil continually. In Genesis 6, God gave mankind 120 years to repent before He sent the great flood. Because of this, we see that God is willing to change His path of coming judgment. The Great Tribulation will come, but it doesn't have to be our generation that ushers it.

We humans are copies of a once-perfect creation. We are a more corrupt version of what once was and we are in need of a savior. That savior is Jesus Christ.

To end this work, I'll leave you with two quotes:

Bill Nye:

> Science is the key to our future, and if you don't believe in science, then you're holding everybody back. And it's fine if you as an adult want to run around pretending or claiming that you don't believe in evolution, but if we educate a generation of people who don't believe in science, that's a recipe for disaster.

God:

> **Romans 1:22,** Professing themselves to be wise, they became fools...